T0181336

Studies in Systems, Decision and Control

Volume 113

Series editor

Janusz Kacprzyk, Polish Academy of Sciences, Warsaw, Poland
e-mail: kacprzyk@ibspan.waw.pl

About this Series

The series "Studies in Systems, Decision and Control" (SSDC) covers both new developments and advances, as well as the state of the art, in the various areas of broadly perceived systems, decision making and control- quickly, up to date and with a high quality. The intent is to cover the theory, applications, and perspectives on the state of the art and future developments relevant to systems, decision making, control, complex processes and related areas, as embedded in the fields of engineering, computer science, physics, economics, social and life sciences, as well as the paradigms and methodologies behind them. The series contains monographs, textbooks, lecture notes and edited volumes in systems, decision making and control spanning the areas of Cyber-Physical Systems, Autonomous Systems, Sensor Networks, Control Systems, Energy Systems, Automotive Systems, Biological Systems, Vehicular Networking and Connected Vehicles, Aerospace Systems, Automation, Manufacturing, Smart Grids, Nonlinear Systems, Power Systems, Robotics, Social Systems, Economic Systems and other. Of particular value to both the contributors and the readership are the short publication timeframe and the world-wide distribution and exposure which enable both a wide and rapid dissemination of research output.

More information about this series at http://www.springer.com/series/13304

Alexander Barkalov · Larysa Titarenko
Jacek Bieganowski

Logic Synthesis for Finite State Machines Based on Linear Chains of States

Foundations, Recent Developments and Challenges

 Springer

Alexander Barkalov
Institute of Metrology, Electronics
  and Computer Science
University of Zielona Góra
Zielona Góra
Poland

Jacek Bieganowski
Institute of Metrology, Electronics
  and Computer Science
University of Zielona Góra
Zielona Góra
Poland

Larysa Titarenko
Institute of Metrology, Electronics
  and Computer Science
University of Zielona Góra
Zielona Góra
Poland

ISSN 2198-4182   ISSN 2198-4190   (electronic)
Studies in Systems, Decision and Control
ISBN 978-3-319-86714-4   ISBN 978-3-319-59837-6   (eBook)
DOI 10.1007/978-3-319-59837-6

Printed on acid-free paper

This Springer imprint is published by Springer Nature
The registered company is Springer International Publishing AG
The registered company address is: Gewerbestrasse 11, 6330 Cham, Switzerland

Contents

Abbreviations

BCM	Block of collections of microoperations
BCT	Block of code transformer
BIMF	Block of input memory functions
BIT	Block of inputs transformer
BM	Benchmark
BMO	Block of microoperations
BMTS	Block of transformation of states into microoperations
BOT	Block of output transformation
BRAM	Block of random access memory
BRLC	Block of replacement of logical conditions
BTMS	Block of transformation of microoperations into states
CAD	Computer-aided design
CLB	Configurable logic block
CM	Control memory
CMCU	Compositional microprogram control unit
CMO	Collection of microoperations
CT	Counter
ECMO	Extended collection of microoperations
EFSM	FSM based on elementary linear chains of states
ELCS	Elementary linear chains of states
EMB	Embedded memory block
EMBer	Logic block consisting from EMBs
FF	Flip-flop
FSM	Finite state machine
FPGA	Field-programmable gate arrays
GFT	Generalized formula of transition
GSA	Graph-scheme of algorithm
HFPGA	Hybrid field-programmable gate array
IOB	Input–output block
LCC	Linear chain of classes

LCS	Linear chain of states
LUT	Look-up table
LUTer	Logic block consisting from LUTs
MCU	Microprogram control unit
MPI	Matrix of programmable interconnections
MX	Multiplexer
NFSM	FSM based on normal linear chains of states
NLCS	Normal linear chains of states
PAL	Programmable array logic
PEO	Pseudoequivalent outputs
PES	Pseudoequivalent states
PLA	Programmable logic arrays
PLAer	Logic block consisting from PLAs
PROM	Programmable read-only memory
RG	Register
RLC	Replacement of logical conditions
ROM	Read-only memory
SFT	System of formulae of transitions
SG	System gate
SOP	Sum of products
ST	Structure table
STG	State transition graph
UFSM	FSM based on unitary linear chains of states
ULCS	Unitary linear chains of states
XFSM	FSM based on extended linear chains of states
XLCS	Extended linear chains of states

Chapter 1
Introduction

Now we witness the very rapid development of computer science. Computers and embedded systems can be found in practically all fields of human activity. The up-to-day state of the art in this area is characterized by three major factors. The first factor is a development of ultra complex VLSI such as "system-on-programmable chip" (SoPC) with billions of transistors and hundreds of millions of equivalent gates [11]. The second factor is a development of hardware description languages (HDL) such as VHDL and Verilog [3, 6, 7] that permits to capture a design with tremendous complexness. The third factor is a wide application of different computer-aided design (CAD) tools to design very complex projects in the satisfactory time [10, 12, 16]. These three factors affected significantly the process of hardware design. Now the hardware design is very similar to the development of computer programs. An application of HDLs together with CAD-tools allows concentrating the designer's energy on the basic problems of design, whereas a routine work remains the prerogative of computers.

Tremendous achievements in the area of semiconductor electronics turn micro-electronics into nanoelectronics. Actually, we observe a real technical boom connected with achievements in nanoelectronics. It results in development of very complex integrated circuits, particularly in the field of programmable logic devices. Our book targets field-programmable gate arrays (FPGA) [13, 14]. Up-to-day FPGAs have up to 7 billion of transistors [15]. So, they are so huge, that it is enough only one chip to implement a very complex digital system including a data-path and a control unit. Because of the extreme complexity of modern microchips, it is very important to develop effective design methods targeting particular properties of logical elements in use.

As it is known, any digital system can be represented as a composition of a data-path and a control unit [2]. Logic circuits of operational blocks forming a data-path have regular structures [1]. It allows using standard library elements of CAD tools (such as counters, multibit adders, multipliers, multiplexers, decoders and so on) for

© Springer International Publishing AG 2018
A. Barkalov et al., *Logic Synthesis for Finite State Machines Based on Linear Chains of States*, Studies in Systems, Decision and Control 113,
DOI 10.1007/978-3-319-59837-6_1

their design. A control unit coordinates interplay of other system blocks producing a sequence of control signals. These control signals cause executing some operations in a data-path. As a rule, control units have irregular structures. It makes process of their design very sophisticated. In the case of complex logic controllers, the problem of system design is reduced practically to the design of control units [2]. Many important features of a digital system, such as performance, power consumption and so on, depend to a large extent on characteristics of its control unit. Therefore, to design competitive digital systems with FPGAs, a designer should have fundamental knowledge in the area of logic synthesis and optimization of logic circuits of control units. As experience of many scientists shows, design methods used by standard industrial packages are far from optimal [8]. Especially it is true in the case of designing complex control units. It means that a designer could be forced to develop his own design methods, next to program them and at last to combine them with standard packages to get a result with desired characteristics. To help such a designer, this book is devoted to solution of the problems of logic synthesis and reduction of hardware amount in control units. We discuss a case when a control unit is represented by the model of finite state machine (FSM). The book contains some original synthesis and optimization methods based on the taking into account the peculiarities of a control algorithm and an FSM model in use. Regular parts of these models can be implemented using such library elements as embedded memory blocks, decoders and multiplexers. It results in reducing the irregular part of the control units described by means of Boolean functions. It permits decreasing for the total number of look-up table (LUT) elements in comparison with logic circuits based on known models of FSM. Also, it makes the problem of place-and-routing much simpler. The third benefit is the reducing power dissipation in comparison with FSM circuits implemented only with LUTs. In our book, control algorithms are represented by graph-schemes of algorithms (GSA) [3]. This choice is based on obvious fact that this specification provides the simple explanation of the methods proposed by the authors.

To minimize the number of LUTs in FSM logic circuits, we propose to replace a state register by a state counter. Such replacement is executed in the case of compositional microprogram control units [4]. But those methods are based on creating some linear sequences (chains) of ioperator vertices where only unconditional interstate transitions are possible. We propose an approach allowing creating linear chains of states. Such chains can have more than one output (and more than one input). It simplifies the system of input memory functions and, therefore, decreases the number of LUTs in the resulting FSM circuit. We combine this approach with using EMBs for implementing the system of output functions (microoperations). It allows a significant decreasing for the number of LUTs, as well as eliminating a lot of interconnections in the FSM logic circuit. It saves area occupied by the circuit and diminishes the resulting power dissipation. Of course, it leads to more sophisticated synthesis process than the one connected only with using LUTs.

The process of FSM logic synthesis is reduced to a transformation of a control algorithm into some tables describing the behaviour of FSM blocks [5, 9]. These tables are used to find the systems of Boolean functions, which can be used to implement logic circuits of particular FSM blocks. In order to implement corresponding circuits, this information should be transformed using data formats of particular industrial CAD systems. We do not discuss this step is in our book. Our book contains a lot of example showing design of FSMs with using the proposed methods. Some examples are illustrated by logic circuits. The main part of the book contains seven chapters.

Chapter 2 provides some basic information. Firstly, the language of GSA is introduced. Next, the connections are shown with GSAs and state transition graphs of both Mealy and Moore FSMs. Classical principles of FSM logic synthesis are discussed. The basic features of FPGA are analyzed. It is shown that embedded memory blocks allow implementing systems of regular Boolean functions. The modern design flow is analyzed targeting FPGA-based projects. Next, the basic problems of FSM design are considered. Different state assignment methods are analyzed, as well as the methods of functional decomposition. Next the issues are discussed connected with implementing FSM logic circuits with EMBs. The peculiarities of hybrid FPGAs are discussed in last part of the Chapter.

Chapter 3 is devoted to the using linear chains in FSMs. The counter-based microprogram control units are discussed, as well as known PLA-based structures of Moore FSMs. Then, there are discussed methods of optimal state assignment and transformation of state codes into codes of classes of pseudoequivalent states (PES). Next, there are introduced different linear chains of states (LCS) such as unitary, elementary, normal and extended LCSs. The structural diagrams are proposed for LCS-based Moore FSMs. The proposed procedures are discussed for constructing different linear chains of states.

Chapter 4 is devoted to the problems of hardware reducing for FPGA-based logic circuits of Moore FSMs. The design methods are proposed based on using more than one source of codes of classes of pseudoequivalent states (PES). Two structural diagrams and design methods are proposed for Moore FSM based on transformation of objects. The first method is based on transformation the unitary codes of microoperations into the codes of PES. The second approach is connected with transformation of the codes of collections of microoperations into the codes of PES. The last part of the Chapter is devoted to the replacement of logical conditions.

Chapter 5 deals with optimization of logic circuits of hybrid FPGA-based Mealy FSMs. First of all, the models with two state registers are discussed. This approach allows removal of direct dependence among logical conditions and output functions of Mealy FSM. Next, the proposed design methods are presented. Some improvements are proposed for further hardware reduction. They are based on the special state assignment and transformation of state codes. The proposed methods target joint using such blocks as LUTs, PLAs and EMBs in FSM circuits. The models are discussed based on the principle of object transformation. The last part of the chapter is connected with design methods connected with the object transformation.

Chapter 6 is devoted to hardware reduction targeting the elementary LCS-based Moore FSMs. Firstly, the optimization methods are proposed for the base model of EFSM. They are based on the executing either optimal state assignment or transformation of state codes. Two different models are proposed for the case of code transformation. They depend on the numbers of microoperations of FSM and outputs of EMB in use. The models are discussed based on the principle of code sharing. In this case, the state code is represented as a concatenation of the chain code and the code of component inside this chain. The last part of the chapter is devoted to design methods targeting the hybrid FPGAs.

Chapter 7 is devoted to hardware reduction targeting the normal LCS-based Moore FSMs. Firstly, the optimization methods are proposed for the base model of NFSM. They are based on the executing either optimal state assignment or transformation of state codes. Two different models are proposed for the case of code transformation. They depend on the numbers of microoperations of FSM and outputs of EMB in use. The models are discussed based on the principle of code sharing. In this case, the state code is represented as a concatenation of the code of normal LCS and the code of component inside this chain. The last part of the chapter is devoted to design methods targeting the hybrid FPGAs.

Chapter 8 is devoted to hardware reduction targeting the extended LCS-based Moore FSMs. Firstly, the design method is proposed for the base model of XFSM. Next, the methods are proposed targeting the hardware reduction in the circuits based on this model. They are based on the executing either optimal state assignment or transformation of state codes. The third part deals with the models based on the encoding of the chain outputs. At last, the principle of code sharing is discussed. In this case, the state code is represented as a concatenation of the code of class of pseudoequivalent chains and the code of component inside this class.

We hope that our book will be interesting and useful for students and PhD students in the area of Computer Science, as well as for designers of modern digital systems. We think that proposed FSM models enlarge the class of models applied for implementation of control units with modern FPGA chips.

References

1. Adamski, M., Barkalov, A.: Architectural and Sequential Synthesis of Digital Devices. University of Zielona Góra Press (2006)
2. Baranov, S.: Logic Synthesis for Control Automata. Kluwer Academic Publisher (1994)
3. Baranov, S.I.: Logic and System Design of Digital Systems. TUT press Tallinn (2008)
4. Barkalov, A., Titarenko, L.: Logic Synthesis for Compositional Microprogram Control Units. vol. 22, Springer (2008)
5. Barkalov, A., Titarenko, L.: Logic Synthesis for FSM-based Control Units. Springer (2009)
6. Bergmann, R.: Experience Management: Foundations, Development Methodology, and Internet-based Applications. Springer (2002)
7. Ciletti, M.D.: Advanced Digital Design with the Verilog HDL. Prentice Hall (2011)
8. Czerwinski, R., Kania, D.: Finite State Machine Logic Synthesis for Complex Programmable Logic Devices. vol. 231 Springer (2013)
9. De Micheli, G.: Synthesis and Optimization of Digital Circuits. McGraw-Hill Higher Education (1994)
10. Grout, I.: Digital Systems Design with FPGAs and CPLDs. Elsevier (2011)
11. Hamblen, J.O., Hall, T.S., Furman, M.D.: Rapid Prototyping of Digital Systems: SOPC Edition. vol. 1 Springer (2007)

12. Jenkins, J.H.: Designing with FPGAs and CPLDs. Prentice Hall, New York (1994)
13. Maxfield, C.: The Design Warrior's Guide to FPGAs: Devices. Elsevier, Tools and Flows (2004)
14. Maxfield, C.: FPGAs: Instant Access. Elsevier (2011)
15. Skliarova, I., Skliarov, V., Sudnitson, A.: Design of FPGA-based Circuits Using Hierarchical Finite State Machines. TUT Press, Tallinn (2012)
16. Zeidman, B., Designing with FPGAs and CPLDs. CRC Press (2002)

Chapter 2
Finite State Machines and Field-Programmable Gate Arrays

Abstract The Chapter provides some basic information. Firstly, the language of GSA is introduced. Next, the connections are shown with GSAs and state transition graphs of both Mealy and Moore FSMs. Classical principles of FSM logic synthesis are discussed. The basic features of FPGA are analyzed. It is shown that embedded memory blocks allow implementing systems of regular Boolean functions. The modern design flow is analyzed targeting FPGA-based projects. Next, the basic problems of FSM design are considered. Different state assignment methods are analyzed, as well as the methods of functional decomposition. Next the issues are discussed connected with implementing FSM logic circuits with EMBs. The peculiarities of hybrid FPGAs are discussed last part of the Chapter.

2.1 Background of Finite State Machines

Finite state machines (FSM) are the most widely used components of digital systems [7, 15, 45]. In this book, we use FSMs for representing and synthesis of control units [8]. To represent a control algorithm, the language of graph-schemes of algorithms (GSA) is used in our book [6]. This language gives the better understanding of ideas discussed in this book.

A graph-scheme of algorithm Γ is a directed connected graph having finite set of vertices. There are four different types of vertices (Fig. 2.1): start, end, operator and conditional.

The start vertex has no input; it corresponds to the beginning of a control algorithm. The end vertex has no output; it corresponds to the finishing of a control algorithm. An operator vertex contains a collection of output signals executed in a particular cycle of a digital system's operation. Let us call this collection $Y_t \subseteq Y$ a collection of microoperations (CMO). The set Y includes microoperations of a digital system: $Y = \{y_1, \ldots, y_N\}$. A conditional vertex includes an input variable $x_l \in X$ checked for branching a control algorithm. So, the set $X = \{x_1, \ldots, x_L\}$ is a set of logical conditions.

© Springer International Publishing AG 2018   7
A. Barkalov et al., *Logic Synthesis for Finite State Machines Based on Linear Chains of States*, Studies in Systems, Decision and Control 113,
DOI 10.1007/978-3-319-59837-6_2

Fig. 2.1 Types of vertices of
GSA

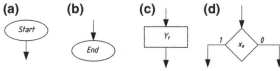

Fig. 2.2 Graph-scheme of
algorithm Γ_1

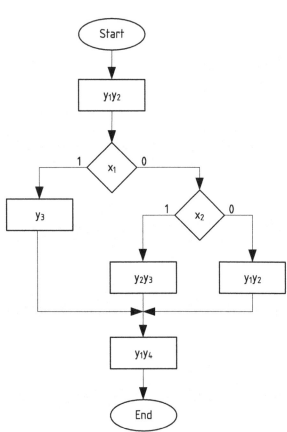

Let us analyse the GSA Γ_1 (Fig. 2.2). The following sets and their characteristics
can be found for this GSA: the set of microoperations $Y = \{y_1, \ldots, y_4\}$ having $N = 4$
elements and the set of logical conditions $X = \{x_1, x_2\}$ having $L = 2$ elements.

The following collections of microoperations can be derived from the operator
vertices of $\Gamma_1 : Y_1 = \{y_1, y_2\}, Y_2 = \{y_3\}, Y_3 = \{y_2, y_3\}, Y_4 = \{y_1, y_4\}$. So, there are
$T_o = 4$ different CMOs in the discussed case. Let us point out that different operator
vertices could include the same CMOs. Also, different conditional vertices could
include the same logical conditions.

A control unit generates a sequence of CMOs distributed in time. To start the
execution of a control algorithm, the special pulse Start is used. There are the different
sequences of CMOs for GSA Γ_1 (Fig. 2.3). They depend on the values of logical

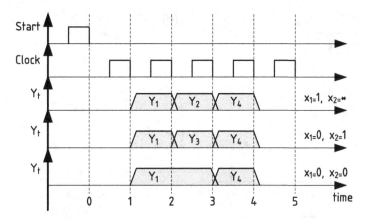

Fig. 2.3 Operation of control unit represented by GSA Γ_1

conditions $x_l \in X$. The record $x_2 = *$ means that the value of x_2 does not affect the outputs of a control unit.

So, to produce a sequence of CMOs, it is necessary to have some information about the prehistory of system operation. The prehistory for the instant t is determined by input signals $X(0), X(1), \ldots, X(t-1)$ in the previous time intervals. It means that an output signal $Y(t)$ is determined by the following function:

$$Y(t) = f(X(0), \ldots, X(t-1), X(t)). \qquad (2.1)$$

To represent the prehistory, interrenal states of FSM are used [6]. They form a set $A = \{a_1, \ldots, a_M\}$. In any instant t ($t = 1, 2, \ldots$) an FSM could be in some state $a_m \in A$. As a rule, if $t = 0$, then an FSM is in the initial state $a_1 \in A$.

An FSM could be represented by the following vector

$$S = \langle A, X, Y, \gamma, \lambda, a_1 \rangle. \qquad (2.2)$$

In (2.2), the function γ determines either conditional or unconditional transitions $\langle a_m, a_s \rangle$, where $a_m, a_s \in A$. The function λ determines outputs of FSM. There are two basic models of FSM, namely Moore FSM and Mealy FSM. For both models, the function γ is determined as the following:

$$a(t+1) = \gamma(a(t), x(t)). \qquad (2.3)$$

In the case of Moore FSM, the function λ is determined as:

$$y(t) = \lambda(a(t)). \qquad (2.4)$$

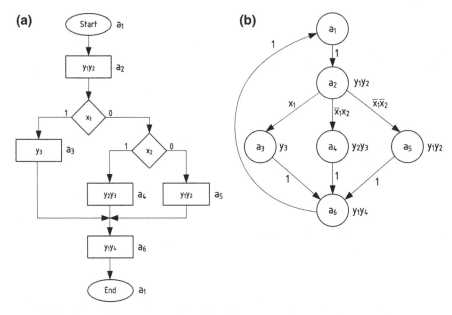

Fig. 2.4 Marked GSA Γ_1 (**a**) and STG of Moore FSM (**b**)

So, the outputs of a Moore FSM depend on its internal states. The outputs of a Mealy FSM depend on both inputs and states:

$$y(t) = \lambda(a(t), x(t)). \qquad (2.5)$$

The difference in functions (2.4) and (2.5) follows from different approach used for constructing the set A [6]. In the case of Moore FSM, each operator vertex is marked by a unique state $a_m \in A$. Both start and end vertices are marked by the initial state $a_1 \in A$. The marked GSA Γ_1 and corresponding state transition graph (STG) of Moore FSM are shown in Fig. 2.4.

The arcs of STG correspond to transitions between the FSM's states. For unconditional transitions, the arcs are marked by "1". For conditional transitions, the arcs are marked by conjunctions of input variables causing these transitions. As follows form Fig. 2.4, the Moore FSM has $M = 6$ states: $A = \{a_1, \ldots, a_6\}$.

The following rules are used for finding the states of Mealy FSM [6]. The input of end vertex is marked by the initial state $a_1 \in A$, as well as the output of the start vertex. If some vertex is connected with the output of an operator vertex, then its input is marked by a unique state. Each input can be marked only once. Using this procedure to GSA Γ_1, the following marked GSA and STG can be obtained (Fig. 2.5).

The arcs of STG of a Mealy FSM are marked by pairs ⟨input variables, output variables⟩. As follows from Fig. 2.5, the Mealy FSM has $M_o = 3$ states: $A = \{a_1, a_2, a_3\}$.

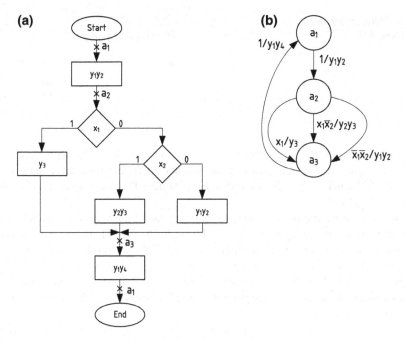

Fig. 2.5 Marked GSA Γ_1 (**a**) and STG of Moore FSM (**b**)

The FSMs shown in Figs. 2.4 and 2.5 are equivalent, because they are constructed for the same GSA Γ_1. Comparison of these FSMs leads to the following conclusion. For equivalent Mealy and Moore FSMs the following relations take places:

$$M_o \leq M; \tag{2.6}$$
$$H_o \leq H. \tag{2.7}$$

In (2.7) the symbol $H_o(H)$ stands the number of transitions (the number of arcs of STG) of Mealy (Moore) FSM.

2.2   Synthesis of Mealy and Moore FSMs

Let us start from the synthesis of Mealy FSM. Let us construct the set of states A for some GSA Γ_j. This set includes M_o states. Let us encode each state $a_m \in A$ by a binary code $K(a_m)$ having R_o bits:

$$R_O = \lceil \log_2 M_O \rceil. \tag{2.8}$$

Fig. 2.6 Structural diagram
of P Mealy FSM

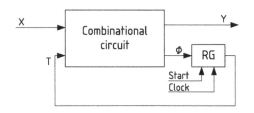

So, the value of R_o is determined as a ceil function [32]. The expression $\lceil B \rceil$ determines the least integer greater than or equal to B. Let us use the state variables $T_r \in T$ where $|T| = R_o$ for encoding of states. This step is named the state assignment [15].

There are many methods of state assignment targeting mostly optimization of hardware amount in the logic circuit of an FSM [4, 15, 17, 51]. There are thousands of publications connected with the state assignment. We do not discuss them in this chapter.

The Mealy FSM can be represented as P FSM (Fig. 2.6). It is a composition of a combinational circuit (CC) and register RG. The CC implements the system of input memory functions

$$\Phi = \Phi(T, X). \tag{2.9}$$

The set $\Phi = \{D_1, \ldots, D_{R_o}\}$ includes input memory functions used for changing the content of RG. The RG includes R_o of D flip-flops. A flip-flop number r represents the state variable $T_r (r = 1, \ldots, R_o)$. Also, the CC implements the system of output variables

$$Y = Y(T, X). \tag{2.10}$$

The pulse *Start* loads the code $K(a_1)$ of the initial state $a_1 \in A$ into RG. As a rule, this code includes all zeros [6]. The pulse *Clock* allows changing the content of RG determined by input memory functions (2.9). Let us point out that the system (2.9) determines the function (2.3), whereas the system (2.10) the function (2.5).

The method of Mealy FSM synthesis on the base of a GSA Γ includes the following steps [6]:

1. Construction of marked GSA and finding the set of states A.
2. State assignment.
3. Constructing the structure table of FSM.
4. Deriving systems Φ and Y from the structure table.
5. Implementing FSM logic circuit using some logic elements.

In this book, we use the symbol $P(\Gamma_j)$ to show that the model of P Mealy FSM is synthesized using the GSA Γ_j. Let us discuss an example of synthesis for Mealy FSM $P(\Gamma_1)$

Table 2.1 Structure table of Mealy FSM $P(\Gamma_1)$

a_m	$K(a_m)$	a_s	$K(a_s)$	X_h	Y_h	Φ_h	h
a_1	00	a_2	01	1	$y_1 y_2$	D_2	1
a_2	01	a_3	10	x_1	y_3	D_1	2
		a_3	10	$\bar{x}_1 x_2$	$y_2 y_3$	D_1	3
		a_3	10	$\bar{x}_1 \bar{x}_2$	$y_1 y_2$	D_1	4
a_3	10	a_1	00	1	y_4	–	5

The marked GSA Γ_1 is shown in Fig. 2.5a. The following set of states A exists in this case: $A = \{a_1, a_2, a_3\}$. It means that $M_O = 3$, $R_O = 2$, $T = \{T_1, T_2\}$ and $\Phi = \{D_1, D_2\}$. Let us endode the states $a_m \in A$ in the trivial way: $K(a_1) = 00$, $K(a_2) = 01$ and $K(a_3) = 10$. An FSM structure table (ST) can be viewed as a list of interstate transitions obtained from STG. This table includes the following columns [6]: a_m is an current state of FSM; $K(a_m)$ is a code of the current state; a_s is a state of transition (next state); $K(a_s)$ is a code of the state $a_s \in A$; X_h is an input signal determining the transition $\langle a_m, a_s \rangle$ and it is equal to the conjunction of some elements (or their complements) of the set X; Y_h is a collection of microoperations generated during the transition $\langle a_m, a_s \rangle$; Φ_h is a set of input memory functions equal to 1 to load the code $K(a_s)$ into RG; h is a number of transition ($h = 1, \ldots, H_O$).

There are $H_O = 5$ arcs in the STG (Fig. 2.5b). Therefore, the ST of FSM $P(\Gamma_1)$ includes $H_O = 5$ rows (Table 2.1).

The connection between the STG (Fig. 2.5b) and the ST (Table 2.1) is obvious. It is clear that this table can be constructed using only the marked GSA Γ_1 (Fig. 2.5a).

Functions Φ and Y are derived from the ST as the sums-of-products (SOP) depending on the following product terms:

$$F_h = A_m X_h \ (h = 1, \ldots, H_O). \tag{2.11}$$

In (2.11), the member A_m is a conjunction of state variables $T_r \in T$ corresponding to the code $K(a_m)$ from the row number h of ST:

$$A_m = \bigwedge_{r=1}^{R_O} T_r^{l_{mr}} \ (m = 1, \ldots, M_O). \tag{2.12}$$

In (2.12), the variable $l_{mr} \in \{0, 1\}$ is a value of the bit r of the code $K(a_m)$ and $T_r^0 = \bar{T}_r$, $T_r^1 = T_r (r = 1, \ldots, R_O)$.

Fig. 2.7 Structural diagram
of PY Moore FSM

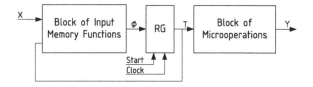

The systems (2.9)–(2.10) are represented by the following SOPs:

$$D_r = \bigvee_{h=1}^{H_O} C_{rh} F_h \ (r = 1, \ldots, R_O); \tag{2.13}$$

$$y_h = \bigvee_{h=1}^{H_O} C_{nh} F_h \ (n = 1, \ldots, N). \tag{2.14}$$

In these expressions, $C_{rh}(C_{nh})$ is a Boolean variable equal to 1 if and only if (iff) the h-th row of ST includes the variable $D_r(y_n)$.

The following equations can be derived from Table 2.1: $F_1 = \bar{T}_1 \bar{T}_2$, $F_2 = \bar{T}_1 T_2 x_1$; $F_3 = \bar{T}_1 T_2 \bar{x}_1 x_2$; $F_4 = \bar{T}_1 T_2 \bar{x}_1 \bar{x}_2$; $F_5 = T_1 \bar{T}_2$; $D_1 = F_2 \vee F_3 \vee F_4$; $D_2 = F_1$; $y_1 = F_1 \vee F_4$; $y_2 = F_1 \vee F_3 \vee F_4$; $y_3 = F_2 \vee F_3$; $y_4 = F_5$. These functions can be minimized. For example, $D_1 = \bar{T}_1 T_2$, $y_1 = \bar{T}_2 \vee \bar{T}_1 T_2 \bar{x}_1 \bar{x}_2$. But we do not discuss this step in our book.

The last step of the discussed method depends on logic elements used for implementing the FSM logic circuit. In this book, we discuss the design methods connected with field-programmable gate arrays (FPGA). We discuss these methods a bit later.

The method of Moore FSM synthesis includes the same steps as for Mealy FSM. Let us discuss the structural diagram of PY Moore FSM (Fig. 2.7).

A block of input memory functions (BIMF) generates the functions $D_r \in \Phi$ represented by the system (2.9). In the case of Moore FSM, the minimum number of bits required for the state assignment is determined as

$$R = \lceil \log_2 M \rceil. \tag{2.15}$$

A block of microoperations (BMO) generates the mirooperations $y_n \in Y$, where

$$Y = Y(T). \tag{2.16}$$

The Eq. (2.16) follows from (2.4). So, the outputs of Moore FSM depend only on its states. Due to this property, the relations (2.6)–(2.7) take places.

In the formula PY, the letter "P" shows that a structure diagram includes the BIMF. The letter "Y" means that there is the BMO in the FSM's structural diagram. Of course, both blocks can be combined into a single combinational circuit as it is for the Mealy FSM (Fig. 2.6). But we want to show that systems Φ and Y have the different nature. These systems are based on different product terms.

Table 2.2 Structure table of Moore FSM PY(Γ_1)

a_m	$K(a_m)$	a_s	$K(a_s)$	X_h	Φ_h	h
a_1	000	a_2	001	1	D_3	1
$a_2(y_1 y_2)$	001	a_3	010	x_1	–	2
		a_4	011	$\bar{x}_1 x_2$	$D_2 D_3$	3
		a_5	100	$\bar{x}_1 \bar{x}_2$	D_1	4
$a_3(y_3)$	010	a_6	101	1	$D_1 D_2$	5
$a_4(y_2 y_3)$	011	a_6	101	1	$D_1 D_2$	6
$a_5(y_1 y_2)$	100	a_7	101	1	$D_1 D_2$	7
$a_6(y_1 y_4)$	101	a_1	000	1	–	8

Let us discuss an example of synthesis for the Moore FSM PY(Γ_1). As follows from Fig. 2.4b, there is $M = 6$, therefore, $R = 3$. Let us encode the states $a_m \in A$ in the trivial way: $K(a_1) = 000, \ldots, K(a_6) = 101$.

The structure table of Moore FSM includes all columns presented in its counterpart of Mealy FSM but the column Y_h. The collections of microoperations are written it the column a_m[6]. The STG (Fig. 2.4b) contains $H = 8$ arcs. Therefore, the ST of Moore FSM PY(Γ_1) includes $H = 8$ rows (Table 2.2).

The connection between Table 2.2 and the STG (Fig. 2.4b) is obvious. The system (2.9) depends on the terms (2.11). So, each function $D_r \in \Phi$ is represented by SOP (2.13). Of course, the symbol H_O should be replaced by H. The functions $y_n \in Y$ depend on the terms (2.12). Of course, the symbols R_O and M_O in (2.12) should be replaced by R and M, respectively. So, the BMO implements the following functions:

$$y_n = \bigvee_{m=1}^{M} C_{nm} A_m \ (n = 1, \ldots, N). \tag{2.17}$$

In (2.17), the Boolean variable $C_{nm} = 1$, iff the microoperation $y_n \in Y$ is placed in the column a_m of the ST.

The following terms and functions, for example, can be derived from Table 2.2:
$F_1 = \bar{T}_1 \bar{T}_2 \bar{T}_3$; $F_2 = \bar{T}_1 \bar{T}_2 T_3 x_1$; $A_1 = \bar{T}_1 \bar{T}_2 \bar{T}_3$; $F_1 = \bar{T}_1 \bar{T}_2 \bar{T}_3$; $D_1 = F_4 \vee F_5 \vee F_6 \vee F_7$; $y_1 = A_2 \vee A_5 \vee A_6$.

Let us point out that the system (2.16) can be represented by a truth table [9]. In the discussed case it is Table 2.3.

Analysis of Table 2.3 shows that system Y for the discussed case is determined for more than 50% of possible input assignments. It takes place for any Moore FSM if exactly R bits are used the state assignment. Such functions are named regular [8]. The best way for implementing logic circuits for regular functions is using memory blocks [8]. These blocks could be either read-only memories (ROM) or random access memories (RAM). We discuss those approaches a bit further.

Table 2.3 Truth table for system Y

$K(a_m)$	Y_t	m
$T_1 T_2 T_3$	$y_1 y_2 y_3 y_4$	
000	0000	1
001	1100	2
010	0010	3
011	0110	4
100	1100	5
101	1001	6
110	****	*
111	****	*

2.3 Field-Programmable Gate Arrays

Field-programmable gate arrays were invented by designers of Xilinx in 1984 [29]. Their influence on different directions of engineering has been growing extremely fast. One of the most important reasons for this process is a relatively cheap development cost. These chips can replace billions 2NAND gates (system gates) [30]. The first FPGAs were used for implementing simple and glue logic [45]. Now they have up to 7 billions transistors [45], posses clock frequency acceding gigahertz, their the most advanced technology is 22 nm [45].

The world's first FPGA XC2064 (Xilinx, 1985) offered 85 000 transistors, 128 logic cells, 64 configurable logic blocks (CLB) based on three-input look-up table (LUT) elements having clock frequency up to 50 MHz. In accordance with [45], from 1990 to 2005 FPGA grew 200 times in capacity, became 40 times faster, 500 times cheaper, reduced power consumption in 50 times. Analysis conducted by the authors of [45] shows that from 2005 till 2011 the capacity of FPGA has been increased in at least 10 times. Five companies dominate on the FPGA market: Xilinx, Altera, Lattice Semiconductor, Microsemi and QuickLogic. All their products can be found on corresponding homepages [2, 28, 31, 36, 52].

In this Chapter we discuss only the basic features of FPGAs relevant to implementing logic circuits of control units. Let us analyze peculiarities of LUT-based FPGAs. As a rule, typical FPGAs include four main elements: configurable logic blocks based on LUTs, matrix of programmable interconnections (MPI), input-output blocks (IOB) and embedded memory blocks (EMB). The organization of an FPGA chip is shown in Fig. 2.8.

As a rule, LUTs are based on RAM having limited amount of inputs S ($S \leq 6$). A single LUT can implement an arbitrary Boolean function depended on L input variables ($L \leq S$) represented by a truth table.

A typical CLB includes a single LUT, programmable flip-flop (FF), multiplexer (MX) and logic of clock and set-reset (LCSR). The simplified structure of CLB is shown in Fig. 2.9.

Fig. 2.8 Simplified
organization of FPGA

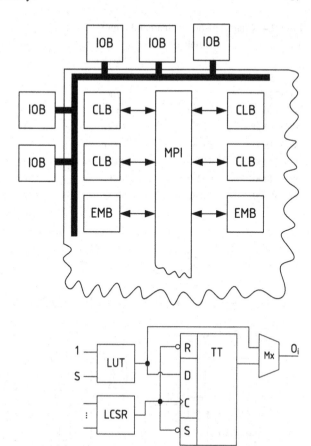

Fig. 2.9 Simplified structure
of CLB

The output of LUT is connected with FF which could be programmed as D, JK, or T flip-flop. The FF could be by-passed due to programmable MX. So, the output O_i of a CLB can be either combinational or registered. The existence of flip-flops allows organization of either registers or counters. Both these devices are used for FSM implementation.

To show the progress in FPGA characteristics, let us start from the family Spartan-3 by Xilinx [28]. They were introduced in 2002, were powered by 1, 2 V and used the 90 nm technology. They included LUTs having 4 inputs. The chips of Spartan-3 included up to 104 EMBs with 18Kb for each of them. These blocks are named block of RAMs (BRAM). So, the chips included up to 1,87 Mb of BRAMS. The frequency of operation for these FPGAs was variable (from 25 MHz till 325 Mhz). Some characteristics of Spartan-3 family are shown in Table 2.4.

The second column of Table 2.4 contains the number of system gates (SG) for a chip. The column 4 determines the capacity of memory created by LUTs. It is named distributed random-access memory (DRAM).

Table 2.4 Characteristics of Spartan-3 family

Device	Number		Capacity in bits	
	CLB	SG (K)	BRAMs (K)	DRAM (K)
XC3550	1728	50	72	12
XC35200	4320	200	216	30
XC35400	8064	400	288	56
XC351000	17280	1000	432	120
XC351500	29952	1500	576	208
XC352000	46080	2000	720	320
XC354000	62208	4000	1728	432
XC355000	74880	5000	1872	520

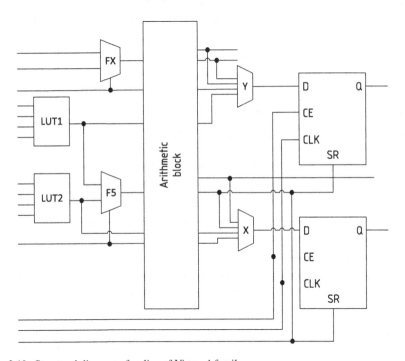

Fig. 2.10 Structural diagram of a slice of Virtex-4 family

The structure of CLB has become more and more complex with the development of technology. For example, the CLB of Virtex-7 includes 4 slices having fast interconnections. A slice includes 2 LUTs, four multiplexers, arithmetic logic and two programmable flip-flops (Fig. 2.10).

This slice includes 2 LUTs; each of them has $S = 4$ inputs. Each LUT can implement an arbitrary logic function depended on 4 variables. Using the multiplexer F5, both LUTs are viewed as a single LUT having $S = 5$. The multiplexer FX combines

Table 2.5 Characteristics of FPGAs by Xilinx

Family	Modification	Number of slices	Capacity in Kbits		Technology nm
			BRAMs	DRAM	
Virtex-4	LX	10 752–89 088	1 296–6 048	168–1 392	90
	SX	10 240–24 576	2 304–5 760	160–384	
	FX	5 472–63 168	648–9 936	86–987	
Virtex-5	LX	4 800–51 840	1 152–10 368	320–3 420	65
	LXT	3 120–51 840	936–11 664	210–3 420	
	SXT	5 440–37 440	3 024–18 576	520–4 200	
	TXT	17 280–24 320	8 208–11 664	1 500–2 400	
	FXT	5 120–30 720	2 448–16 416	390–2 280	
Virtex-6	LXT	11 640–118 560	5 616–25 920	1 045–8 280	40
	SXT	49 200–74 400	25 344–28 304	5 090–7 640	
	HXT	39 360–88 560	18 144–32 832	3 040–6 370	
	CHT	11 640–37 680	5 616–14 976	1 045–3 650	
Virtex-7	T	44 700–305 400	14 760–46 512	3 475–21 550	28
	XT	64 400–135 000	31 680–64 800	6 525–13 275	
	HT	45 000–135 000	21 600–64 800	4 425–13 275	

together outputs of F5 and FX from other slides. So, a slice can implement a Boolean functions depending on 5 variables; two slices on 6 variables; four slices (a CLB) on 7 variables. The arithmetic block allows organizing adders and multiplexers. Multiplexers Y and X determine input data for programmable flip-flops. So, each CLB can include either RG or CT.

The number of inputs per a LUT is increased up to 5 for Virtex-5 family, whereas CLBs of Virtex-6 and Virtex-7 include LUTs having $S = 6$. There are different modifications of FPGAs for each family. We do not discuss them. Some characteristics of modern FPGA chips by Xilinx are shown in Table 2.5.

Analysis of Tables 2.4 and 2.5 proves out statement about the tremendous progress in FPGAs. Let us point out that modern chips include blocks of digital signal processors and central processing units. But these blocks are not used for FSM design. So, we do not discuss them.

As it follows from Table 2.5, modern FPGA includes huge blocks of memory. Let us name these blocks embedded-memory blocks (EMB). EMBs have a property of configurability. It means that they have the constant size (V_O) but both the numbers of cells (V) and their outputs (t_F) can be changeable. There are the following typical configurations of EMBs: $36 K \times 1$, $18 K \times 2$, $8 K \times 4$, $4 K \times 8$ ($4 K \times 9$), $2 K \times 16$ ($2 K \times 18$), $1 K \times 32$ ($1 K \times 36$) and $512 \times ,64$ (512×72) bits [2, 28, 31, 36, 52].

Let an EMB contain V cells having t_F outputs. Let V_O be a number of cells if there is $t_F = 1$. So, the number of V can be determined as

T₁ \ T₂T₃	00	01	11	10
0	0	$y_1 y_2$	$y_2 y_3$	y_3
1	$y_1 y_2$	$y_1 y_4$	*	*

Fig. 2.11 Karnaugh map for system Y

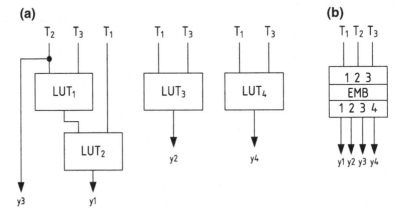

Fig. 2.12 Logic circuit for system (2.19)

$$V = \left\lceil \frac{V_O}{t_F} \right\rceil. \qquad (2.18)$$

Let us point out that decreasing t_F by 1 leads to doubling number of cells (and vice versa).

Embedded-memory bloc Ks could be used for implementing regular functions [8]. Let us discus the Karnaugh map (Fig. 2.10b) corresponding to Table 2.3.

The following functions can be found from this map:

$$y_1 = T_1 \vee \bar{T}_2 T_3;$$
$$y_2 = \bar{T}_1 T_3 \vee T_1 \bar{T}_3; \qquad (2.19)$$
$$y_3 = T_2;$$
$$y_4 = T_1 T_3.$$

Let us use system gates (LUTs with $S = 2$) for implementing the system (2.19). It leads to the logic circuit having 4 LUTs and 2 levels of LUTs (Fig. 2.12a).

From Fig. 2.11a the following negative features can be seen:

1. Different propagation time for different output functions,
2. Input variables should be connected with more than one logic element.

The second feature (bigger value of fan-out for inputs T_1-T_3) also leads to more complicated routing process.

If an EMB is used for implementing the system (2.19), all these problems are absent (Fig. 2.12b). Each input T_1-T_3 is connected only with a single input of EMB. All functions $y_1, y_2, \ldots, y_4$ have the same propagation time. This example is very simple. But a single EMBs having the configuration 512×64 could replace at least 64 LUTs. It is possible if a system Y depend on 9 inputs and includes up to 64 different functions. Of course, the circuit includes 64 LUTs having $S = 9$. But there is no such LUTs in modern FPGAs. If minimization allows dependance of each function $y_n \in Y$ on 8 variables, then 256 LUTs with $S = 6$ are necessary for creating a logic circuit.

In a typical FPGA 60% of power is consumed by the programmable interconnections, 16% is consumed by programmable logic and 24

Replacement of LUTs by EMBs allows decreasing of the number of interconnections. So, it is very important to use EMBs in implementing FSM circuits.

The exceptional complexity of FPGA requires using computer-aided design (CAD) tools for designing logic circuits [20]. It assumes development of formal methods for synthesis and verification of control units [19, 22, 33, 40]. For example, a design process for FPGAs from Xilinx includes the following steps:

1. **Specification of a project**. A design entry can be executed by the schematic editor (if a design is represented by a circuit), or the state editor (a design in represented by an STG) or a program written with some hardware description languages (HDL). The most popular HDLs are VHDL and Verilog [12, 13]. This initial specification is verified and corrected if necessary.
2. **Logic synthesis**. During this step, the package FPGA Express executes synthesis and optimization of an FSM logic circuit. As an outcome of this step, an FPGA Netlist file is generated. This file is represented in either EDIF or XNF format. During this step, library cells from system and user libraries are used.
3. **Simulation**. The functional correctness of an FSM is checked. This step is executed without taking into account real propagation times in a chip. If the outcome of simulation is negative, then the previous steps should be repeated.
4. **Implementation of logic circuit**. Now the Netlist is translated into an internal format of CAD system. Such physical objects as CLBs and chip pins are assigned for initial Netlist elements. This step is named the packing. The step of mapping is the first stage of the packing. The mapping refers to the process of associating entities such as gate-level functions in the gate-level netlist with the LUT-level functions available on the FPGA [29]. It is not a one-to-one mapping because each LUT can be used to represent a number of logic gates [21]. The mapping step gives results for executing the packing. During this step, the LUTs and flip-flops are packed into the CLBs. Both mapping and packing steps are very

difficult because there are many variants of their solutions. Following packing the step of place-and-route is executed. Now we know the connections between CLBs and parts of logic functions are implemented. But there are many ways how these CLBs could be placed in the FPGA. The placement problem is also very difficult because hundreds of thousands or even millions CLB should be placed. During the routing, it is necessary to decide how to connect all CLBs for a particular project. This step should be executed in a way giving the maximum possible performance. Obviously, the outcome of placement affects tremendously the outcome of routing. When routing is finished, the real performance could be found. Also, the BitStream is formed which will be used for chip programming.

5. **Project verification**. The final simulation is performed where the actual values of delays among the physical elements of a chip are used. If outcome of this step is negative (the actual performance of an FSM is less than it is necessary), then the previous steps of the design process should be repeated.

6. **Chip programming**. This step is connected with the writing of the final bit stream into the chip.

One of time most important roles in the design process plays the step of logic synthesis. Let us analyze this step for FGPA-based FSMs.

2.4   Implementing FSMs with FPGAs

The synthesis is a transformation of initial specification of project into the structural specification where elements of lower abstraction levels are used [1]. The synthesis process is repeated till each element to be assigned is represented by some library element. In the case of FSM with FPGAs, the library elements are LUTs and EMBs.

An FSM circuit includes LUTs and flip-flops. To get a structure of FSM, the sequential synthesis is executed. It transforms specifications of FSM (GSA, STG) into structure tables describing some parts of an FSM logic circuit. Next, the systems of Boolean functions are derived from those tables. These systems could be (2.9), (2.10) or (2.16). The stage of logic synthesis follows the sequential synthesis. Now, the functions are transformed into smaller subsystems. Each of these subsystems could be implemented using either a LUT or an EMB of a particular FPGA chip. Both these steps are considered in our book. We combine them in a single stage of synthesis of FSM logic circuit.

If an FSM is specified by a GSA, then such sets as X, Y and A are known. But there are no state codes. To obtain them, the step of state assignment is executed [15]. This step is very important because its outcome has a tremendous influence on the hardware amount (the number of LUTs) in the FSM logic circuit [1]. A strategy of state assignment could target optimization for area, performance, power consumption or testability.

One of the most popular state assignment algorithms is JEDI which is distributed with the system SIS [43]. JEDI targets a multi-level logic implementation. It is based on the weight assignment for state.

The input dominant algorithm assigns higher weights to pairs of present states which asserts similar inputs and produce sets of next states. It allows maximizing the size of common cubes in the implemented logic function. The output dominanted algorithm assigns higher weights to pairs of next states which are generated by similar input combinations and similar sets of present states. It maximizes the number of common cubes in the logic function.

In modern industrial packages different state assignment strategies are used. For example, two optimization criteria are used in the design tool XST of Xilinx: maximum performance and minimum hardware [53]. Seven different approaches are used for state assignment. The automatic state assignment is based od some special algorithm proposed by Xilinx. It has been never published. The method of one-hot encoding is based on the following expression:

$$R = M. \tag{2.20}$$

This method is very popular because it is very simple and each LUT is connected with a flip-flop. So, this conception is implemented very easy in FPGAs. In this case, there is a lot of input memory functions but each of them is relatively small. The compact state assignment is based on formula (2.8) for a Mealy FSM and the formula (2.15) for a Moore FSM. In this case the number of input memory functions is minimum possible, but they are rather complex. In this book we mostly use this approach and name it a binary state assignment. Two other methods ar based on codes either Gray or Johnson. At last, there are so named speed encoding and the sequential encoding based on using of the counter instead of state register.

The master thesis [48] is devoted to investigation of the influence of state assignment methods on characteristics of Mealy FSM. The benchmarks from [54] are used in the investigation. The results obtained for Mealy FSM are represented in Table 2.6. The efficiency of the investigated methods is shown in Figs. 2.13, 2.14 and 2.15.

The investigations are executed for the FPGA XC5VLX30 of Xilinx. The first column of table Table 2.6 shows the name of a benchmark. The columns "LUT" show number of LUTs in the final circuit. The columns "MHz" represent the maximal frequency of operation for final Mealy FSMs.

The best results are produced when the automatic state assignment is used. It gives the best outcomes for area (58,54% of all benchmarks) and performance (39,02%). The binary state assignment posses the second place in this competition. As follows from Fig. 2.15, the automatic state assignment produces the best results when both area and performance are optimized (29,27%). The same results are produced for the compact (binary) state assignment. It is interesting that the one-hot state assignment can optimize only one parameter of FSM circuit. (Fig. 2.14)

Of course, these results are true only for the chip XC5VLX30 but similar conclusions are made, for example, in [27]. It allows to suggest that these conclusions have a rather common nature.

Table 2.6 Outcomes of investigations

FSM	Auto		One-hot		Compact		Sequential		Gray		Johnson		Speedl	
	LUT	MHz	LUT	MHz	LUT	MHz	LUT	MHz	LUT	MHz	LUT	MHz	LUT	MHz
bbara	11	639	9	966	13	635	13	639	19	589	24	545	13	962
bbsse	29	559	29	559	29	582	29	538	31	538	36	408	38	556
bbtas	5	962	8	966	5	966	5	962	5	955	5	962	9	966
beecount	7	952	19	639	7	952	7	952	7	948	21	625	30	583
cse	49	480	52	477	46	463	50	487	46	454	71	434	72	453
dkl4	8	945	29	522	8	945	8	945	8	945	19	623	40	512
dkl5	7	1062	19	737	7	1062	7	1062	7	1062	7	1062	19	659
dkl6	46	556	46	556	15	625	19	506	27	554	86	355	70	399
dkl7	6	952	14	669	6	952	6	952	6	952	7	895	27	571
dk27	5	900	8	906	5	897	5	959	5	955	6	899	10	903
dk512	17	730	17	730	7	899	7	895	7	899	21	437	19	790
exl	64	586	64	586	74	447	67	478	66	406	106	340	72	605
ex4	15	962	15	962	16	626	15	598	14	748	33	546	15	962
ex6	29	553	30	580	20	621	23	615	22	616	36	426	31	598
keyb	56	384	56	384	65	358	71	382	66	447	62	435	85	374
kirkman	51	874	84	1058	53	569	48	880	51	874	112	451	84	1058
lion	3	1084	5	962	3	1080		1080	3	1084	3	1084	5	962
markl	27	726	27	726	19	708	22	622	18	623	27	574	29	959
mc	5	1071	8	1071	5	1071	6	1071	5	1071	5	1071	8	1071
opus	22	596	22	754	21	628	26	585	22	596	26	576	26	671
planet	100	888	100	888	138	389	145	417	149	375	192	346	106	637
planetl	100	888	100	888	138	389	145	417	149	375	192	346	106	637
pma	73	554	73	554	115	438	108	367	112	375	121	405	88	559
si	77	550	77	550	75	447	89	328	105	368	114	361	81	552
sl488	140	425	140	425	141	432	130	394	147	433	192	334	162	458
si494	124	412	124	412	143	442	135	383	145	383	192	333	152	462
s208	28	559	28	559	13	669	12	716	15	639	29	483	50	386
s27	4	962	21	636	4	962	7	679	4	962	12	664	21	631
s298	362	406	362	406	330	313	264	311	274	314	716	244	399	397
s386	26	577	31	586	28	581	28	558	29	429	43	422	36	441
s420	28	559	28	559	14	629	12	716	15	639	29	483	36	510
s510	42	900	42	900	39	448	53	440	50	427	123	388	42	900
s820	63	429	63	429	85	395	92	441	93	438	98	366	93	399
s832	63	429	63	429	73	412	77	431	87	394	97	335	108	444
sand	99	569	99	569	121	426	125	421	125	438	189	306	103	490
scf	179	676	179	676	202	338	205	349	197	389	337	327	180	561
shiftreg	0	1584	9	1080	0	1584	4	959	4	959	5	902	4	903
sse	29	559	29	559	28	543	37	548	32	540	44	394	36	612
styr	118	430	118	430	127	369	138	363	138	353	181	323	161	454
tav	6	1556	6	1556	6	911	6	911	5	914	5	914	6	1556
tbk	55	406	179	360	71	465	129	342	137	290	295	276	444	342

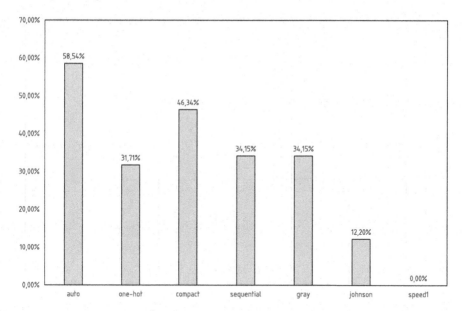

Fig. 2.13 Efficiency of state assignment methods for area optimization

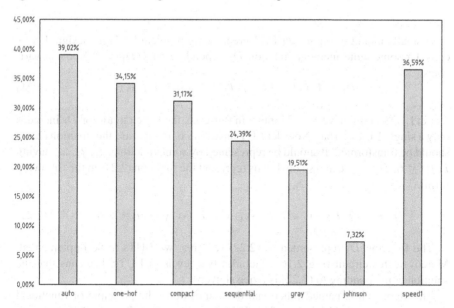

Fig. 2.14 Efficiency of state assignment methods for performance optimization

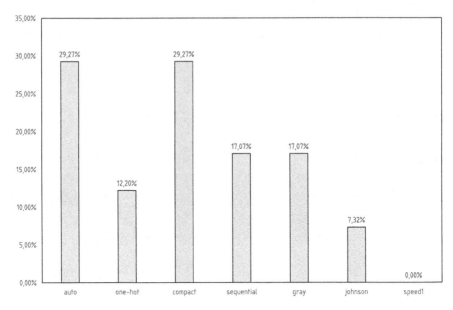

Fig. 2.15 Efficiency of state assignment methods for area/performance optimization

A small amount of inputs per LUT create a big problem for logic design. Let us consider some input memory function D_1 depending on $I(D_1) = 7$ Boolean variables:

$$D_1 = T_1\bar{T}_2 T_3 x_1 \bar{x}_2 \vee T_1 T_2 T_3 x_3 x_4 \vee T_1 \bar{T}_2 \bar{T}_3 x_1 x_3. \tag{2.21}$$

Let LUTs in use have $S = 7$ inputs. In this case, the logic circuit for D_1 includes only a single LUT (2.16a). Now, let LUT have $S = 6$. In this case the function (2.21) should be transformed. It should be represented by some functions $f_1, f_2, \ldots$ having $I(f_1) \leq 6, I(f_2) \leq 6$ and so on. Let us represent the function (2.21) in the following form:

$$D_1 = T_1(\bar{T}_2 T_3 x_1 \bar{x}_2 \vee T_2 T_3 x_3 x_4) \vee \bar{T}_1 \bar{T}_2 T_3 \bar{x}_1 x_3 = T_1 A \vee B. \tag{2.22}$$

The function D_1 represented as (2.22) requires two LUTs to be implemented. Moreover, this circuit (Fig. 2.16b) includes two levels of LUTs. It means that the solution corresponding to (2.22) is twice slower.

This approach is named functional decomposition. The principle of functional decomposition is the basic one for FPGA-based design [37, 41]. This approach usually targets only LUTs, but also there are methods using EMBs as tools for implementing some subfunctions [11, 38].

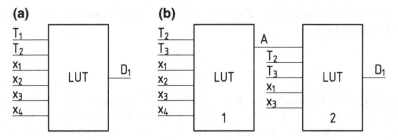

Fig. 2.16 Implementing function D_1 with LUTs having $S = 7$ (**a**) and S=6 (**b**)

Fig. 2.17 Illustration of the principle of functional decomposition

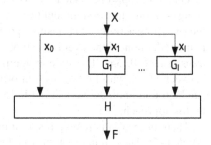

In general, the method of functional decomposition is based on representation of a Boolean function $F(X)$ in the following form:

$$F(X) = H(X_0, G_1(X_1), \ldots, G_I(X_I)). \tag{2.23}$$

The Eq. (2.23) corresponds to the implementation of the circuit shown in Fig. 2.17.

The negative influence of functional decomposition is increasing of the propagation time in comparison with single-level circuit. It follows from complain [23], the methods of functional decomposition are far from ideal. Let us point out that is is very important to decrease the number of arguments and product terms in Boolean functions to be implemented. We discus these methods a bit further.

Modern FPGAs posses the substantial logic resources and high processing speeds. Due to these factors, FPGAs now are used for some applications previously targeted to Application Specific Integrated Circuits (ASIC) [49]. Now FPGAs are used in portable computing devices and wireless telecommunication equipment. They are also used extensively in space-based applications [31]. The rising of FPGA complexity leads to increasing the power consumed by FPGA-based devices. It is known that FSMs consume a significant amount of power in any FPGA-based project [49]. Therefore, minimizing power consumed by the FSMs can significantly reduce the total power consumed by a device.

The dynamic power dissipated in CMOS circuits can be represented by the well-known formula [47]:

$$P = \sum_{n=1}^{N} C_n f_n V_{DD}^2. \tag{2.24}$$

In (2.24), N is the number of elements, C_n is the load capacitance at the output of the element number n, f_n is the frequency of its switching, and V_{DD} is the supply voltage. One of the ways for decreasing the power dissipation is decreasing of the switching activity of flip-flops [47].

One of the approaches leading to decreasing the power dissipation in FSMs is the energy-saving state assignment [39]. Main works in low-power FSMs compute first the switching activity and transition probabilities [50]. The key idea of these methods is the reduction of the average activity by minimizing the bit changes during state transitions [10, 34]. The state assignment should minimize the Hamming distance between states with high transition probability. Different variants of this approach can be found in many works [5, 14, 16, 35]. There are hundreds of articles devoted to this approach.

There is a very interesting result of investigations conducted by the authors of the article [47]. They found that the smaller FSM circuits consumes less power than its bigger versions. It is clear because a smaller circuit needs less interconnections than its bigger counterpart. One of the ways leading to smaller FSM circuit is application of EMBs for implementing some parts of FSM circuits [46]. It is shown that FSM implementation with EMBs provides some benefits compared to synthesis with LUTs [18, 42]. The maximum clock frequency of an FSM implemented in a ROM block is independent of its complexity. Of course, it is possible if the whole circuit is implemented using just a single EMB. The memory blocks of FPGAs provide control signals that allow for module deactivation when the FSM is inactive. It provides an efficient mechanism for power saving. It has been proved [49] that complex FSMs consume less power when implemented as memory blocks. Let us consider some EMB-based models of FSMs.

In the simplest case, it is enough a single EMB for implementing an FSM logic circuit. Our book is mostly devoted to Moore FSMs. Because of it, let us discuss possible trivial models of EMB-based Moore FSMs. Analysis of design methods from [18, 47, 49] allows finding four EMB-based models of Moore FSM (Fig. 2.18).

In the simplest case, the P Moore FSM is used (Fig. 2.18a). To use this model, the following condition should take place:

$$2^{L+R}(R + N) \leq V_0. \tag{2.25}$$

To design the logic circuit of Moore FSM $P(\Gamma_j)$, the initial structure table should be transformed. Each row of the transformed ST corresponds to a single cell of EMB. This table includes $V(P)$ rows, where

$$V(P) = 2^{L+R}. \tag{2.26}$$

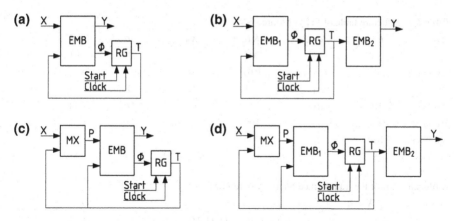

Fig. 2.18 Models of Moore FSMs based on RAMs

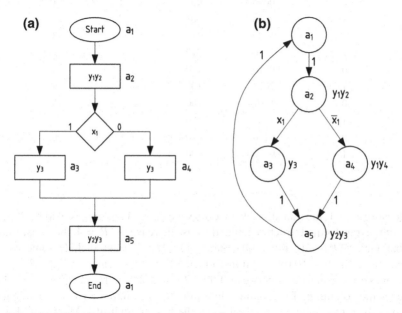

Fig. 2.19 Initial GSA Γ_2 (a) and state transition graph (b)

It is necessary to have 2^L rows for representing transitions from the state $a_m \in A$. The transformed ST includes the following columns: $K(a_m)$, X, $Y(a_m)$, Φ, v. First two columns form the address of a cell. The column $Y(a_m)$ includes a collection of microoperations $Y_t \subseteq Y$ generated in the state $a_m \in A$. The column v contains the numbers of rows (or cells).

Let us consider the initial GSA Γ_2 (Fig. 2.19a). The corresponding state transition graph of Moore FSM is shown in Fig. 2.19b. The structure table of P Moore FSM is represented by Table 2.7.

Table 2.7 Structure table of P(Γ_2) Moore FSM

a_m	$K(a_m)$	a_s	$K(a_s)$	X_h	Φ_h	h
a_1	000	a_2	001	1	D_3	1
$a_2(y_1 y_2)$	001	a_3	010	x_1	D_2	2
		a_4	011	$\bar{x}_1$	$D_2 D_3$	3
$a_3(y_3)$	010	a_5	100	1	D_1	4
$a_4(y_1 y_4)$	011	a_5	100	1	D_1	5
$a_5(y_2 y_3)$	100	a_1	000	1	–	6

Table 2.8 Part of the transformed ST for Moore FSM $P(\Gamma_2)$

$K(a_m)$	X	$Y(a_m)$	Φ	v	h
$T_1 T_2 T_3$	x_1	$y_1 y_2 y_3 y_4$	$D_1 D_2 D_3$		
000	0	0000	001	1	1
000	1	0000	001	2	1
001	0	1100	011	3	3
001	1	1100	010	4	2
010	0	0010	100	5	4
010	1	0010	100	6	4
011	0	1001	100	7	5
011	1	1001	100	8	5
100	0	0110	000	9	6
100	1	0110	000	10	6

Because $L = 1$, the transitions from each state $a_m \in A$ are represented by $2^L = 2$ rows of the transformed structure table. Because there is $L + R = 4$, the transformed ST includes $V(P) = 16$ rows. Of course only $M \cdot 2^L$ rows include some useful information. The first 10 rows of transformed ST are represented by Table 2.8

To make the connection between Tables 2.7 and 2.8 more transparent, the last includes the column h. This column shows the rows of initial ST corresponding to the rows of transformed ST. To implement the logic circuit of P Moore FSM, it is enough to load the bit-stream corresponding to Table 2.8 into a particular EMB.

If condition (2.25) is violated, then other models can be used. Let the following conditions take places:

$$R \cdot 2^{L+R} = V_0; \tag{2.27}$$

$$N \cdot 2^R = V_0. \tag{2.28}$$

In this case, the model of PY Moore FSM (Fig. 2.18b) can be used. In this model, the EMB_1 implements the circuit of BIMF, the EMB_2 implements the circuit of BMO.

Table 2.9 Table of microoperations of Moore FSM PY(Γ_1)

$K(a_m)$	$Y(a_m)$	m
$T_1 T_2 T_3$	$y_1 y_2 y_3 y_4$	
000	0000	1
001	1100	2
010	0010	3
011	1001	4
100	0110	5

To design PY FSM, the table of microoperations should be constructed. It includes the columns $K(a_m)$, $Y(a_m)$, m. In the case of Moore FSM PY(Γ_2), this table includes 8 rows but only $M = 5$ of them include some useful information (Table 2.9). The transformed ST of PY Moore FSM does not include the column $Y(a_m)$.

Let us point out that conditions (2.25) or (2.27) have places only for very simple FSMs. If they are violated, the models based on the replacement of logical conditions [6] are used. These models are represented by Fig. 2.18c, d.

The replacement of logical conditions (RLC) is reduced to replacement of the set X by a set of additional variables $P = \{p_1, \ldots, p_G\}$, where $G \ll L$. As it is stated in [46], there is $G \leq 3$ for a vast majority of practical control algorithms. As a rule, the value of G is determined as $\max(L_1, \ldots, L_M)$. Here the symbol L_m stands for the number of elements in the set $X(a_m) \subseteq X$. These conditions determine transitions from state $a_m \in A$. Let us denote the model (Fig. 2.18c) as MP Moore FSM, the model (Fig. 2.18d) as MPY Moore FSM.

In these FSMs, the multiplexer MX generates functions

$$P = P(T, X). \tag{2.29}$$

The system of input memory functions is represented as

$$\Phi = \Phi(T, P). \tag{2.30}$$

Let the condition (2.25) be violated, but the following condition take place:

$$2^{G+R}(R + N) \leq V_0. \tag{2.31}$$

In this case the MP FSM is used. If the condition (2.31) is violated, then the MPY FSM can be used if the following condition takes place

$$R \cdot 2^{G+R} \leq V_0. \tag{2.32}$$

Of course, the condition (2.28) also should take place.

In this book, we use such terms as LUTer and EMBer. The LUTer is a network of LUTs implementing circuit of BIMF. The EMBer is a network of EMBs imple-

Fig. 2.20 Structural diagram of FPGA-based PY Moore FSM

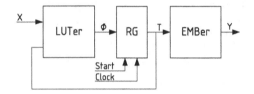

menting the circuit of BMO. The register RG is distributed among the CLBs of the LUTer. For example, the PY Moore FSM (Fig. 2.7) can be represented as the following structure (Fig. 2.20).

There is the extension of the library [54] named 1993 MCNC library of benchmarks (BM). It includes 190 BMs for different FSMs. There is the analysis of these BMs presented in [24]. In total, the BMs comprise 36 304 Boolean functions. An examinations of BMs shown that more than 70% of functions depend on more than 4 variables. Also, roughly 20% of the functions have fan-in equal or greater than 6. If the number of arguments exceeds the number of inputs of a LUT, then such a function might be implemented with programmable logic arrays (PLA) [24].

The PLAs were introduced by Signetics in the mid 1970s [8]. The particular property of PLA is the programmability of both AND– and OR– planes. It provides greater flexibility than PLD where only one plane is programmable. As it mentioned in [25], being coupled with LUTs, PLAs provide an integrated programmable resource that can be used in many digital systems design to support control logic for LUT-based data-paths.

Finite state machines are wide fan-in, low logic-density circuits [25]. To optimize the chip area occupied by such circuits, architectures of hybrid FPGAs (HPGA) were proposed [26, 44]. They include LUTs, EMBs and SRAM-configurable programmable logic arrays (PLA). For example, an Embedded System Block of Altera APEX20K can be configured as a PLA with 32 inputs, 32 product terms and 16 outputs [3].

So, the main programmable logic blocks can be found in FPGAs, namely, LUTs, EMBs and PLAs. In this book, we use the name PLAer for a network of PLAs implementing some part of an FSM circuit. Now let us discuss in details the design methods based on linear chains of states (LCS) in FSMs.

References

1. Adamski, M., Barkalov, A.: Architectural and Sequential Synthesis of Digital Devices. University of Zielona Góra Press (2006)
2. Altera Corp. Altera Homepage. http://www.altera.com
3. Altera Corporation. APEX20K PLD Family Data Sheet (2004)
4. Ashar, P., Devadas, S., Newton. A. R.: Sequential Logic Synthesis. Springer (1992)
5. Bacchetta, P., Daldoss, L., Sciuto, D., Silvano. C.: Low-power state assignment techniques for finite state machines. In: The 2000 IEEE International Symposium on Circuits and Systems, 2000. Proceedings ISCAS 2000 Geneva, vol. 2, pp. 641–644. IEEE (2000)

6. Baranov, S.I.: Logic Synthesis for Control Automata. Kluwer Academic Publishers (1994)
7. Baranov. S.I.: Logic and System Design of Digital Systems. TUT press Tallinn (2008)
8. Barkalov, A., Titarenko. L.: Logic Synthesis for FSM-based Control Units. Springer (2009)
9. Barkalov, A., Titarenko, L.: Basic Principles of Logic Design. UZ Press, Zielona Góra (2010)
10. Benini, L., De Micheli, G.: State assignment for low power dissipation. IEEE J. Solid State Circuits **30**(3), 258–268 (1995)
11. Borowik, G.: Synthesis of sequential devices into FPGA with embedded memory blocks
12. Brown, S., Vranesic, Z.: Fundamentals of Digital Logic with VHDL Design. McGraw-Hill, New York (2000)
13. Brown, S., Vranesic, Z.: Fundamentals of Digital Logic with Verilog Design. McGraw-Hill (2003)
14. Chen, C., Zhao, J., Ahmadi, M.: A semi-gray encoding algorithm for low-power state assignment. In: Proceedings of the 2003 International Symposium on Circuits and Systems, 2003. ISCAS'03, vol. 5, pp. 389–392. IEEE (2003)
15. De Micheli, G.: Synthesis and Optimization of Digital Circuits. McGraw-Hill Higher Education (1994)
16. El-Maleh, A., Sait, S.M., Khan, F.N.: Finite state machine state assignment for area and power minimization. In: 2006 IEEE International Symposium on Circuits and Systems, pp. 5303–5306. IEEE (2006)
17. Eschermann, B.: State assignment for hardwired VLSI control units. ACM Comput. Surv. (CSUR) **25**(4), 415–436 (1993)
18. Garcia-Vargas, I., Senhadji-Navarro, R., Jimenez-Moreno, G., Civit-Balcells, A., Guerra-Gutierrez, P.: ROM-based finite state machine implementation in low cost FPGAs. In: 2007 IEEE International Symposium on Industrial Electronics, pp. 2342–2347. IEEE (2007)
19. Hachtel, G.D., Somenzi, F.: Logic Synthesis and Verification Algorithms. Kluwer Academic Publishers (2000)
20. Jenkins, J.H.: Designing with FPGAs and CPLDs. Prentice Hall, New York (1994)
21. Kahng, A.B., Lienig, J., Igor, L.M., Hu, J.: VLSI Physical Design: From Graph Partitioning to Timing Closure. Springer (2011)
22. Kam, T., Villa, T., Brayton, R.K., Sangiovanni-Vincentelli, A.: Synthesis of Finite State Machines: Functional Optimization. Springer (2013)
23. Kania, D.: Programmable Logic Devices. PWN, Warsaw (2012). (in polish)
24. Kaviani, A., Brown, S.: The hybrid field-programmable architecture. IEEE Des. Test Comput. **16**(2), 74–83 (1999)
25. Krishnamoorthy, S., Swaminathan, S., Tessier, R.: Area-optimized technology mapping for hybrid FPGAs. In: International Workshop on Field Programmable Logic and Applications, pp. 181–190. Springer (2000)
26. Krishnamoorthy, S., Tessier, R.: Technology mapping algorithms for hybrid FPGAs containing lookup tables and PLAs. IEEE Trans. Comput. Aided Des. Integ. Circuits Syst. **22**(5):545–559 (2003)
27. Kubátová, H.: Finite state machine implementation in FPGAs. In: Design of Embedded Control Systems, pp. 175–184. Springer (2005)
28. Lattice Semiconductor. Lattice Semiconductor Homepage. http://www.latticesemi.com
29. Maxfield, C.: The Design Warrior's Guide to FPGAs: Devices, Tools and Flows. Elsevier (2004)
30. Maxfield, C.: FPGAs: Instant Access. Elsevier (2011)
31. Microsemi Corp. Microsemi Homepage. http://www.microsemi.com
32. Minns, P., Elliott, I.: FSM-based digital design using Verilog HDL. Wiley (2008)
33. Navabi, Z.: Embedded Core Design with FPGAs. McGraw-Hill Professional (2006)
34. Nöth, W., Kolla, R.: Spanning tree based state encoding for low power dissipation. In: Proceedings of theConference on Design, Automation and Test in Europe, pp. 168–174. ACM (1999)
35. Park, S.,Cho, S., Yang, S., Ciesielski, M.: A new state assignment technique for testing and low power. In: Proceedings of the 41st annual Design Automation Conference, pp. 510–513, ACM (2004)

36. QuickLogic Corp. Quick Logic. http://www.quicklogic.com
37. Rawski, M., Łuba, T., Jachna, Z., Tomaszewicz, P.: The influence of functional decomposition on modern digital design process. In: Des. Embed. Control Syst. 193–204 (2005). (Springer)
38. Rawski, M., Selvaraj, H., Łuba, T.: An application of functional decomposition in ROM-based FSM implementation in FPGA devices. J. Syst. Archit. **51**(6), 424–434 (2005)
39. Roy, K., Prasad, S.: SYCLOP: Synthesis of CMOS logic for low power applications. In: IEEE 1992 International Conference on Computer Design: VLSI in Computers and Processors, 1992. ICCD'92. Proceedings, pp. 464–467. IEEE (1992)
40. Sasao, T., Brayton, R.K.: Logic Synthesis and Verification. Kluwer Academic Publishers (2002)
41. Scholl, C.: Functional Decomposition with Applications to FPGA Synthesis. Springer (2013)
42. Senhadji-Navarro, R., Garcia-Vargas, I., Jimenez-Moreno, G., Civit-Ballcels, A.: ROM-based FSM implementation using input multiplexing in FPGA devices. Electro. Lett. **40**(20), 1 (2004)
43. Sentovich, E., Singh, K., Lavagno, L., Moon, C., Murgai, R., Saldanha, A., Savoj, H., Stephan, P., Brayton, R., Sangiovanni-Vincentelli, A.: SIS: A System for Sequential Circuit Synthesis. University of California Electronic Research Laboratory, Berkley, Technical report (1992)
44. Singh, S., Singh, R., Bhatia, M.: Performance evaluation of hybrid reconfigurable computing architecture over symmetrical FPGAs. Int. J. Embed. Syst. Appl. **2**(3):107–116 (2012)
45. Skliarova, I., Sklyarov, V., Sudnitson, A.: Design of FPGA-based Circuits Using Hierarchical Finite State Machines. TUT Press, Tallinn (2012)
46. Sklyarov, V.: Synthesis and implementation of RAM-based finite state machines in FPGAs. In: International Workshop on Field Programmable Logic and Applications, pp. 718–727 Springer (2000)
47. Sutter, G., Todorovich, E., López-Buedo, S., Boemo, E.: Low-power FSMs in FPGA: Encoding alternatives. In: International Workshop on Power and Timing Modeling, Optimization and Simulation, pp. 363–370. Springer (2002)
48. Tatolov, T.: Design of FSM with FPGAs. Master's thesis, DonNTU, Donetsk (2011)
49. Tiwari, A., Tomko, K.A.: Saving power by mapping finite-state machines into embedded memory blocks in FPGAs. In: Proceedings of the Conference on Design, Automation and Test in Europe, vol. 2, pp. 916–921. IEEE Computer Society (2004)
50. Tsui, C., Pedram, M., Despain, A.M.: Exact and approximate methods for calculating signal and transition probabilities in FSMs. In: 31st Conference on Design Automation, 1994, pp. 18–23. IEEE (1994)
51. Villa, T., Sangiovanni-Vincentelli, A.: NOVA: State assignment of finite state machines for optimal two-level logic implementations. In: 26th Conference on Design Automation, 1989, pp. 327–332. IEEE (1989)
52. Xilinx Corp. Xilinx Homepage. http://www.xilinx.com
53. Corp, Xilinx: XST User Guide **11**, 3 (2009)
54. Yang, S.: Logic synthesis and optimization benchmarks user guide: version 3.0. technical report, Microelectronics Center of North Carolina (MCNC) (1991)

Chapter 3
Linear Chains in FSMs

Abstract The Chapter is devoted to the using linear chains in FSMs. The counter-based microprogram control units are discussed, as well as known PLA-based structures of Moore FSMs. Then there are discussed methods of optimal state assignment and transformation of state codes into codes of classes of pseudoequivalent states (PES). Next there are introduced different linear chains of states (LCS) such as unitary, elementary, normal and extended LCSs. The structural diagrams are proposed for LCS-based Moore FSMs. The proposed procedures are discussed for constructing different linear chains of states.

3.1 Counter-Based Control Units

As it is pointed out in many works [4, 12], it is necessary to take into account the nature of a control algorithm to optimize characteristics of a control unit. If the set of vertices for a given GSA Γ includes more than 75% of operator vertices, then such a GSA is named linear GSA [8] It means that a corresponding FSM possesses a lot of unconditional transitions. This property can be used for simplifying the system of input memory functions [24]. As a rule, the simplification of input memory functions is possible if the state register is replaced by some more complex device. Two approaches are possible. In the first case a shift register replaces the state register [17, 26]. This approach requires sophisticated algorithms for the state assignment. Because of it, shift register did not find a wide application in FSMs. The second approach is connected with replacement of the state register by a state counter [2, 3, 24, 35]. This approach has found a wide application in the case of microprogram control units [11, 23].

Microprogram control units are based on the operational - address principle for presentation of control words (microinstructions) kept in a special control memory [1]. The typical method of MCU design includes the following steps [8]:

1. Transformation of initial graph-scheme of algorithm.
2. Generation of microinstructions with given format.
3. Microinstruction addressing.
4. Encoding of operational and address parts of microinstructions.

© Springer International Publishing AG 2018   35
A. Barkalov et al., *Logic Synthesis for Finite State Machines Based on Linear Chains of States*, Studies in Systems, Decision and Control 113, DOI 10.1007/978-3-319-59837-6_3

Fig. 3.1 Microinstruction
formats for MCU with
natural addressing of
microinstructions

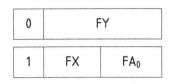

5. Construction of control memory content.
6. Synthesis of logic circuit of MCU using given logical elements.

The mode of microinstruction addressing affects tremendously the method of MCU synthesis [8]. Three particular addressing modes are known:

1. Compulsory addressing of microinstructions.
2. Natural addressing of microinstructions.
3. Combined addressing of microinstructions.

In the first case, the address of microinstruction is kept into a register. This address be viewed as a state code, whereas a microinstruction can be viewed as an FSM state [16]. In the last two cases, the address of microinstruction is kept into a counter.

In the common case, microinstruction formats include the following fields: FY, FX, FA_0 and FA_1. The field FY, operational part of the microinstruction, contains information about microoperations $y_n \in Y (t = 0, 1, \ldots)$, which are executed in cycle t of control unit operation. The field FX contains information about logical condition $x_l^t \in X$, which is checked at time t $(t = 0, 1, \ldots)$. The filed FA_0 contains next microinstruction address A^{t+1} (transition address), either in case of unconditional transition (go to type), or if $x_l^t = 0$. The field FA_1 contains next microinstruction address for the case when $x_l^t = 1$. The fields FX, FA_0 and FA_1 form the address part of microinstruction.

There are two microinstruction formats in case of natural microinstruction addressing [8, 16]: operational microinstructions corresponding to operator vertices of GSA Γ and control microinstructions corresponding to conditional vertices of GSA Γ (Fig. 3.1).

First bit of each format represents field FA, used to recognize the type of microinstruction. Let FA=0 correspond to operational microinstruction and FA=1 to control microinstruction. As follows from Fig. 3.1, next address is not included in operational microinstructions. The same is true for the case, when a logical condition to be checked is equal to 1. In both cases mentioned above current address A^t is used to calculate next address:

$$A^{t+1} = A^t + 1. \tag{3.1}$$

Hence, the following rule is used for next address calculation:

$$A^{t+1} = \begin{cases} A^t + 1 \text{ if } [FA]^t = 0; \\ A^t + 1 \text{ if } (x_l^t = 1) \wedge ([FA]^t = 1); \\ [FA_0]^t \text{ if } (x_l^t = 0) \wedge ([FA]^t = 1); \\ [FA_0]^t \text{ if } ([FX]^t = \varnothing) \wedge ([FA]^t = 1). \end{cases} \tag{3.2}$$

Fig. 3.2 Structural diagram
of MCU with natural
addressing of
microinstructions

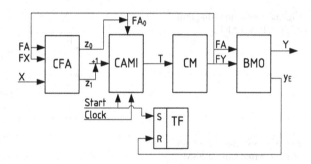

Two first records of (3.2), as well as (3.1), show that a counter should be included into
the structure of MCU with natural addressing. This counter is named a counter of
microinstruction address (CAMI). The structure diagram of MCU includes a block
of addressing (CFA), a control memory (CM), a block of microoperations (BMO)
and a flip-flop TF used for fetching microinstructions from CM (Fig. 3.2).

This MCU operates in the following manner. The pulse *Start* initiates loading
of start address into CAMI. At the same time flip-flop TF is set up. Let an address
A^t be located in CAMI at time t $(t = 0, 1, \ldots)$. If this address determines an
operational microinstruction, the block BMO generates microoperations $y_n \in Y$
and the sequencer CFA produces signal z_1. If this address determines a control
microinstruction, microoperations are not generated, and the sequencer produces
either signal z_0 (corresponding to an address loaded from the field FA_0 or signal z_1
(it corresponds to adding 1 to the content of CAMI). The content of counter CAMI
can be changed by pulse *Clock*. If variable y_E is generated by BMO, then the flip-flop
TF is cleared and operation of MCU terminated.

The MCU (Fig. 3.2) has one serious drawback. Only a single logic condition is
checked during one cycle of MCU's operation. If an GSA includes a lot of multidirectional transitions, the performance of MCU is rather small. But it has very simple
circuit of CFA. It is just a multiplexer. To improve the performance of MCU, the
programmable logic arrays were used for implementing the circuit of CFA.

The PLAs were introduced in the mid 1970s by Signetics [20, 22]. They include
two programmable planes (or arrays), namely, AND-array and OR-array. This property of the PLAs can be used for implementing systems of Boolean functions represented as minimal SOPs [28].

Due to their flexibility, the PLAs found applications in design of control units.
They were used in FSMs to implement both functions Φ and Y [6, 29]. They evoked
a lot of design and optimization methods [29] such as for example, NOVA [36].

In the case of MCU, the block CFA is implemented using a multiplexer [10]. In
articles [30, 31], it was proposed to replace the MX by PLAs. It allowed execution
of multidirectional transitions in a single cycle. Of course, the main drawback of
this approach is the necessity of re-design of CFA if there are some changes in
microprogram to be implemented. The structural diagram of PLA-based MCU is

Fig. 3.3 Structural diagram
of PLA-based MCU

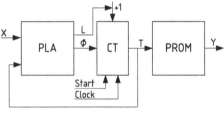

Fig. 3.4 Structural diagram
of PLA-based Moore FSM

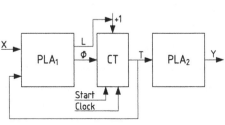

shown in Fig. 3.3. In this model, the control memory is implemented as PROM. The variable L is used for incrementing the counter CT.

If both blocks CFA and CM are implemented with PLAs, it leads to either Mealy or Moore FSM. The structural diagram of PLA-based Moore FSM [2] is shown in Fig. 3.4.

In this structure the PLA_1 implements system (2.9) and function L:

$$L = f(T, X). \tag{3.3}$$

The function L is used for adding 1 to the content of CT. The PLA_2 implements the system of output functions represented as (2.16).

Nowadays, we observe the return of PLAs in logic design. It justifies the statement from [5] that successful computation structures return back at the next round of technological spiral. Now, the return of PLA basis can be observed in the hybrid FPGAs [25, 33], as well as in CoolRunner CPLDs by Xilinx [37].

Also, the PLAs are very popular in the modern sublithographic technology [13, 14]. In the nanoelectronics, these devices are named nano-PLAs. Nowadays, extensive research is conducted in the fields connected with design of different devices based on nano-PLAs [15, 32]. Let us point out that all design methods discussed in this book can be adopted to meet specifics of nano-PLAs.

3.2   Basic Principles of Hardware Reduction for Moore FSMs

Let us consider the structural diagram of Moore FSM (Fig. 3.5).

As it was mentioned before, the block BIMF implements the system of input memory functions (2.9), the block BMO the system of microoperations (2.16).

Fig. 3.5 Structural diagram
of Moore FSM

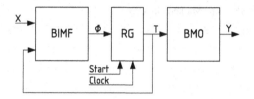

The hardware reduction methods discussed in this section are based on existence of pseudoequivalent states (PES) of Moore FSM [7]. States $a_m, a_k \in A$ are named pseudoequivalent if there are arcs $\langle b_t, b_q \rangle$ and $\langle b_s, b_q \rangle$ where the operator vertice $b_r(b_s)$ is marked by the state $a_m(a_k)$. So, pseudoequivalent states correspond to operator vertices connected with the input of the same vertex of a GSA. The relation of pseudoequivalentness is reflexive, symmetric and transitive. So, it determines some partition $\Pi_A = \{B_1, \ldots, B_I\}$ on the set A. Each element of Π_A is a class of PES.

Let us consider the GSA Γ_3 shown in Fig. 3.6.

The Moore FSM $PY(\Gamma_3)$ has the following parameters: $L = 3$, $N = 6$, $M = 7$, $R = 3$, $H_1 = 15$. So, its structure table includes 15 rows.

As follows from definition of PES, there is the following partition $\Pi_A = \{B_1, \ldots, B_4\}$ for FSM $PY(\Gamma_3)$. Its classes are the following: $B_1 = \{a_1\}$, $B_2 = \{a_2, a_3\}$, $B_3 = \{a_4\}$, $B_4 = \{a_5, a_6, a_7\}$. So, there is $I = 4$.

The simplest way for the hardware reduction is a proper state assignment. There are different approaches named optimal, refined and combined state assignments. Let us discuss these approaches.

In the case of optimal state assignment, the code of each class $B_i \in \Pi_A$ is represented by the minimal possible amount of generalized intervals of R-dimensional Boolean space. The value of R is determined by (2.15). In the best case, each class $B_i \in \Pi_A$ is represented by a single generalized interval. Let us consider the Karnaugh map shown in Fig. 3.7.

It follows from Fig. 3.7 that the class B_1 corresponds to the interval $\langle 0, 0, 0 \rangle$ the class B_2 to the interval $\langle 0, *, 1 \rangle$, the class B_3 to the interval $\langle 0, 1, 0 \rangle$, and the class B_4 to the interval $\langle 1, *, * \rangle$. Each class $B_i \in \Pi_A$ corresponds exactly to a single generalized interval. So, it is the best possible solution. Now, the following class codes can be found: $K(B_1) = 000$, $K(B_2) = 0 * 1$, $K(B_3) = 010$ and $K(B_4) = 1 * *$.

To get the system (2.9), a transformed structure table should be constructed [9]. The table is based on the system of generalized formulae of transitions [9]. In the discussed case, it is the following system:

$$
\begin{aligned}
B_1 &\rightarrow x_1 a_2 \vee \bar{x}_1 a_3; \\
B_2 &\rightarrow x_2 a_4 \vee \bar{x}_2 x_3 a_6 \vee \bar{x}_2 \bar{x}_3 a_3; \\
B_3 &\rightarrow a_5; \\
B_4 &\rightarrow x_3 a_7 \vee \bar{x}_3 a_1.
\end{aligned}
\tag{3.4}
$$

Fig. 3.6 Marked GSA Γ_3

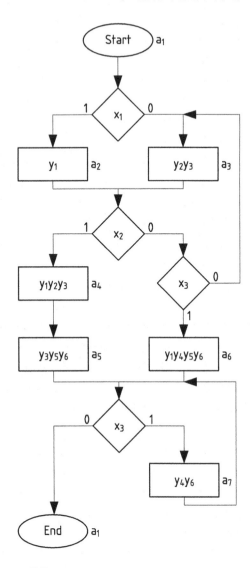

Fig. 3.7 Optimal state codes
of Moore FSM PY(Γ_3)

T_1 \ T_2T_3	00	01	11	10
0	a_1	a_2	a_3	a_4
1	a_5	a_6	*	a_7

Table 3.1 Transformed structure table of Moore FSM $P_0Y(3)$

B_i	$K(B_i)$	a_s	$K(a_s)$	X_h	Φ_h	h
B_1	000	a_2	001	x_1	D_3	1
		a_3	011	$\bar{x}_1$	$D_2 D_3$	2
B_2	0*1	a_4	010	x_2	D_2	3
		a_6	101	$\bar{x}_2 x_3$	$D_2 D_3$	4
		a_3	011	$\bar{x}_2 x_3$	$D_2 D_3$	5
B_3	010	a_5	100	1	D_1	6
B_4	1**	a_7	110	x_3	$D_1 D2$	7
		a_1	000	$\bar{x}_3$	–	8

The connection between the GSA Γ_3 and system (3.4) is obvious. To get such a system, it is necessary:

1. To construct the formulae of transitions for each state $a_m \in A$.
2. To replace a state $a_m \in A$ in the left part of the formula by corresponding class $B_i \in \Pi_A$ such that $a_m \in B_i$.
4. If there are i equal formulae, then only one of them should remain.

We did not show this process in details for constructing the system (3.4).

If the approach of optimal state assignment is used, let us denote such an FSM as P_0Y Moore FSM. The transformed ST of P_0Y Moore FSM includes the following columns: B_i, $K(B_i)$, a_s, $K(a_s)$, X_h, Φ_h, h. In the discussed case, this table includes $H_1 = 8$ rows (Table 3.1).

The connection between Table 3.1 and the system (3.4) is obvious. The state codes are taken from Fig. 3.7. This table is a base for constructing the system (2.9). The system includes the following terms:

$$F_h = \bigwedge_{r=1}^{R} T_i^{l_{hr}} \cdot X_h \ (h = 1, \ldots, H_1). \tag{3.5}$$

The first member of (3.5) represents a conjunction of state variables corresponding to the code $K(B_i)$ of a class B_i from the h-th row of the table. There is $l_{hr} \in \{0, 1, *\}$ and $T_r^0 = \bar{T}_r$, $T_r^1 = T_r$, $T_r^* = 1 \ (r = 1, \ldots, R)$. Let us point out that

$$H_1 \leq H_0 + 1. \tag{3.6}$$

So, the number of rows of transformed ST of Moore FSM is approximately the same as this number for the equivalent Mealy FSM. The input memory functions are represented as the following SOP:

$$D_r = \bigvee_{h=1}^{H_1} C_{rh} F_h \ (r = 1, \ldots, R). \tag{3.7}$$

After minimizing, the following system can be derived from Table 3.1:

$$D_1 = \bar{T}_1 T_3 \bar{x}_2 x_3 \vee \bar{T}_1 T_2 \bar{T}_3 \vee T_1 x_3;$$
$$D_2 = \bar{T}_1 T_3 \vee \bar{T}_1 \bar{T}_2 \bar{T}_3 \bar{x}_1 \vee \bar{T}_1 T_3 \bar{x}_3; \qquad (3.8)$$
$$D_3 = \bar{T}_1 \bar{T}_2 \bar{T}_3 \vee \bar{T}_1 T_3 \bar{x}_2;$$

There are three different approaches for implementing the circuit of PY Moore FSM with FPGAs:

1. LUT-based implementation. In this case both BIMF and BMO are implemented using LUTs. Let us name such a circuit as LFSM.
2. EMB-based implementation. In this case both BIMF and BMO are implemented using EMBs. Let us name such a circuit as MFSM.
3. Heterogeneous implementation. In this case the BIMF is implemented with LUTs, whereas the BMO with EMBs. Let us name such a circuit as HFSM.

In the case of HFSM, the system (3.7) should be minimized [9]. It leads to decreasing the number of LUTs in the circuit of BIMF. In the case of MFSM, there is no need in minimizing [18, 34]. In the case of LFSM both blocks should be optimized.

Let LUT elements in use have S inputs. Let the following condition take place:

$$S \geq R. \qquad (3.9)$$

In this case, any function $y_n \in Y$ is implemented using only a single LUT. If condition (3.8) is violated, then it is necessary to minimize equations from (2.16). It can be executed by applying the refined state assignment [9].

The following system of Boolean equations can be derived from Fig. 3.3:

$$
\begin{aligned}
y_1 &= A_2 \vee A_4 \vee A_6; \\
y_2 &= A_3 \vee A_4; \\
y_3 &= A_3 \vee A_4 \vee A_5; \\
y_4 &= A_6 \vee A_7; \\
y_5 &= A_5 \vee A_6; \\
y_6 &= A_5 \vee A_6 \vee A_7.
\end{aligned}
\qquad (3.10)
$$

Let $S = 2$, it means that condition (3.8) is violated. Let us encode the states $a_m \in A$ as it is shown in Fig. 3.8.

Now, the system (3.10) is transformed into the following one:

Fig. 3.8 Refined state codes for Moore FSM PY(Γ_3)

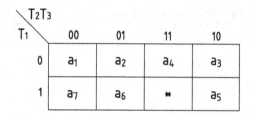

T_1 \ T_2T_3	00	01	11	10
0	a_1	a_2	a_4	a_3
1	a_7	a_6	*	a_5

$$
\begin{aligned}
y_1 &= T_3; \\
y_2 &= \bar{T}_1 T_2; \\
y_3 &= T_2; \\
y_4 &= T_1 \bar{T}_2; \\
y_5 &= T_1 T_3 \vee T_1 T_2; \\
y_6 &= T_1.
\end{aligned}
\tag{3.11}
$$

The system (3.11) results in the logic circuit (Fig. 3.9) having 4 LUTs and two levels of elements.

If the optimal state assignment is used (Fig. 3.7), then the circuit of BMO requires 8 LUTs. Let us point out that the refined state assignment leads to $H_1 = 8$ for the discussed case. But it is just a coincidence.

In the general case, the optimal state assignment optimizes only the circuit of BIMF. In turns, the refined state assignment optimizes only the circuit of BMO. To optimize both blocks of an LFSM, the combined state assignment [9] should be used. This method can be explained as the following. Let us construct the following systems of functions:

$$
Y = Y(A); \tag{3.12}
$$
$$
B = B(A). \tag{3.13}
$$

Let system (3.12) be determined by expression (2.17), while the elements of system (3.13) are represented as

Fig. 3.9 Logic circuit of block BMO

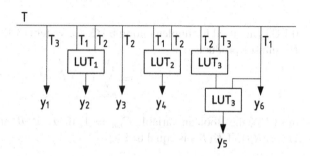

Fig. 3.10 Structural diagram
of $P_C Y$ Moore FSM

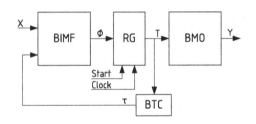

$$B_i = \bigvee_{m=1}^{M} C_{im} A_m \ (i = 1, \ldots, I), \tag{3.14}$$

where C_{im} is a Boolean variable equal to 1, iff $a_m \in B_i$. The combined state assignment is executed in such a manner, that the total number of terms is minimal for systems (3.12) and (3.13). This problem can be solved using, for example, the algorithm JEDI. In the discussed case, the results of combined state assignment are the same as for the refined state assignment (Fig. 3.8). Of course, it is a particular case.

Let us point out that the combined state assignment could produce results which are far from optimal for one or both blocks of PY Moore FSM. In this case the total area can be decreased using a transformer of state codes into codes of the classes of pseudoequivalent states [9]. It results in $P_C Y$ Moore FSM shown in Fig. 3.10.

In $P_C Y$ Moore FSM, a block BIMF implements functions

$$\Phi = \Phi(\tau, X), \tag{3.15}$$

where τ is a set of variables used to code classes $B_i \in \Pi_A$. A code transformer BTC generates codes of classes $B_i \in \Pi_A$ on the base of codes for states $a_m \in B_i$. To encode the classes $B_i \in \Pi_A$, the additional variables $\tau_r \in \tau$ are used. The number of these variables is determined as

$$R_B = \lceil \log_2 I \rceil. \tag{3.16}$$

The block BTC implements the system of Boolean functions

$$\tau = \tau(T). \tag{3.17}$$

If LUTs are used for implementing BTC, the system (3.17) should be represented as R_B sum-of-products:

$$\tau_r = \bigvee_{m=1}^{M} C_{rm} (\bigwedge_{r=1}^{R} A_m^{l_{mr}}). \tag{3.18}$$

In (3.18), the Boolean variable $C_{rm} = 1$, iff $a_m \in B_i$ and the bit number r ($r = 1, \ldots, R_B$) of $K(B_i)$ is equal to 1.

Table 3.2 State transition table of Moore FSM S_1

a_m	a_s	X_h	h
a_1	a_2	x_1	1
$(-)$	a_3	$\bar{x}_1$	2
a_2	a_4	x_2	3
$(y_1 y_3 y_5 y_7)$	a_5	$\bar{x}_2$	4
a_3	a_4	x_2	5
$(y_1 y_4 y_7)$	a_5	$\bar{x}_2$	6
a_4	a_4	x_2	7
(y_3)	a_5	$\bar{x}_2$	8
a_5	a_7	x_3	9
(y_2)	a_6	$\bar{x}_3 x_4$	10
	a_8	$\bar{x}_3 \bar{x}_4$	11
a_6	a_7	x_3	12
$(y_2 y_4 y_7)$	a_6	$\bar{x}_3 x_4$	13
	a_8	$\bar{x}_3 \bar{x}_4$	14
a_7	a_6	x_3	15
$(y_2 y_4 y_7)$	a_7	$\bar{x}_3 x_4$	16
	a_8	$\bar{x}_3 \bar{x}_4$	17
a_8	a_3	x_5	18
(y_6)	a_1	$\bar{x}_5$	19

If EMBs are used for implementing the circuit of BTC, it is represented by a truth table [28]. Let us point out that both BMO and BTC can be implemented using the same EMB.

Consider an example of $P_C Y$ Moore FSM S_1 design, where the FSM is set up by its state transition table (Table 3.2).

The following values and sets can be derived from Table 3.2: $M = 8$, $R = 3$, $\Pi_A = \{B_1, B_2, B_3, B_4\}$, where $B_1 = \{a_1\}$, $B_2 = \{a_2, a_3, a_4\}$, $B_3 = \{a_5, a_6, a_7\}$, $B_4 = \{a_8\}$, $I = 4$. Obviously, there is no such a state assignment variant which gives the transformed structure table with $H_0 = 9$ rows. Remind, this value corresponds to the number of rows in the structure table of the equivalent Mealy FSM. The method of synthesis includes the following steps.

1. **Construction of systems Y and B.** For the Moore FSM S_1, we can construct the functions $y_1 = A_2 \vee A_3$, $y_2 = A_2 \vee A_4 \vee A_7$, $y_5 = A_2 \vee A_7$, $y_6 = A_7 \vee A_8$, $y_7 = A_2 \vee A_3 \vee A_6 \vee A_7$, $B_1 = A_1$, $B_2 = A_2 \vee A_3 \vee A_4$, $B_3 = A_5 \vee A_6 \vee A_7$, $B_4 = A_8$.
2. **State assignment.** For $P_C Y$ Moore FSM, the state encoding targets hardware decrease for block of microoperations. Thus, the refined state encoding should be done. The outcome of this step is shown in Fig. 3.11.

3. **Construction of functions representing the block BMO**. The codes represented by Fig. 3.11 permit to get the following system:

$$
\begin{aligned}
y_1 &= \bar{T}_1 T_3 = \Delta_1; \\
y_2 &= T_1 \bar{T}_2 = \Delta_2; \\
y_3 &= \bar{T}_1 T_2 \vee T_2 T_3 = \Delta_3 \vee \Delta_4; \\
y_4 &= \bar{T}_2 T_3 = \Delta_5; \\
y_5 &= \Delta_4; \\
y_6 &= T_1 T_2 = \Delta_6; \\
y_7 &= T_3 = \Delta_7;
\end{aligned}
\tag{3.19}
$$

4. **Construction of functions representing the block BTC**. Besides, the codes represented by Fig. 3.11 permit to get the following system:

$$
\begin{aligned}
B_1 &= \bar{T}_1 \bar{T}_2 \bar{T}_3 = \Delta_8; \\
B_2 &= \bar{T}_1 T_3 \vee \bar{T}_1 T_2 = \Delta_1 \vee \Delta_3; \\
B_3 &= T_1 \bar{T}_2 \vee T_1 T_3 = \Delta_2 \vee \Delta_{10}; \\
B_4 &= T_1 T_2 \bar{T}_3 = \Delta_9;
\end{aligned}
\tag{3.20}
$$

Because of $I = 4$, there is $R_B = 2$. So, there is a set $\tau = \{\tau_1, \tau_2\}$. Of course, there are a lot of ways for encoding the classes.

Let us encode the classes $B_i \in \Pi_A$ in the trivial way, namely $K(B_1) = 00, \ldots, K(B_4) = 11$. Now we can find that $\tau_1 = B_3 \vee B_4, \tau_2 = B_2 \vee B_4$. It gives the following system of equations:

$$
\begin{aligned}
\tau_1 &= \Delta_2 \vee \Delta_9 \vee \Delta_{10}; \\
\tau_2 &= \Delta_1 \vee \Delta_3 \vee \Delta_9;
\end{aligned}
\tag{3.21}
$$

5. **Construction of transformed structure table**. Let us construct the transformed structure table of the Moore FSM S_1 (Table 3.3). This table includes the following columns: B_i, $K(B_i)$, a_s, $K(a_s)$, X_h, Φ_h, h. For the FSM S_1, the codes $K(B_i)$ can be derived from Fig. 3.11.

The following system Φ is derived from Table 3.3:

$$
\begin{aligned}
D_1 &= F_4 \vee F_5 \vee F_6 \vee F_7; \\
D_2 &= F_1 \vee F_3 \vee F_5 \vee F_7; \\
D_3 &= F_1 \vee F_2 \vee F_5 \vee F_6 \vee F_7.
\end{aligned}
\tag{3.22}
$$

The terms of system Φ are determined as the following conjunctions:

$$
F_h = \bigwedge_{r=1}^{R_0} \tau_r^{l_{hr}} X_h \ (h = 1, \ldots, H_0).
\tag{3.23}
$$

Fig. 3.11 Refined state codes for Moore FSM S_1

T_1 \ T_2T_3	00	01	11	10
0	a_1	a_3	a_2	a_4
1	a_5	a_6	a_7	a_8

Table 3.3 Transformed structure table for P_CY Moore FSM S_1

B_i	$K(B_i)$	a_s	$K(a_s)$	X_h	Φ_h	h
B_1	00	a_2	011	x_1	D_2D_3	1
		a_3	001	$\bar{x}_1$	D_3	2
B_2	01	a_4	010	x_2	D_2	3
		a_5	100	$\bar{x}_2$	D_1	4
B_3	10	a_7	111	x_3	$D_1D_2D_3$	5
		a_6	101	$\bar{x}_3x_4$	D_1D_3	6
		a_8	110	$\bar{x}_3\bar{x}_4$	D_1D_2	7
B_4	11	a_3	001	x_5	D_3	8
		a_1	000	$\bar{x}_5$	–	9

In (3.23), a variable $l_{hr} \in \{0, 1\}$ is equal to the value of the bit r for the code $K(B_i)$, which is written in the row h of the table. From Table 3.3, for example, it can be found that: $F_1 = \bar{\tau}_1\bar{\tau}_2x_1$, $F_2 = \tau_1\bar{\tau}_2\bar{x}_3x_4$ and so on. Let us point out that this table includes $H_0 = 9$ rows, it is the absolute minimum for the Moore FSM S_1.

One of the very popular methods of state assignments is a one-hot state assignment [19]. This method targets FPGA-based implementations of FSMs [21]. It is based on the connection of each LUT with its own flip-flop [27]. But it is shown in [34] that for complex FSMs it is reasonable to use EMBs. Using EMBs presumes application of the binary state assignment based on (2.15). Because all FSMs with counters use EMBs for implementing the system Y, we do not discuss the one-hot approach.

3.3 Linear Chains of States

In this book, we propose the conception of linear chains of states (LCS). Four different kinds of LCS can be found in any GSA Γ. We name them unitary, elementary, normal and extended LCSs. Let us define these types of chains using some part of GSA Γ_0 (Fig. 3.12). Let us introduce some definitions.

Definition 3.1 Each state $a_m \in A$ corresponds to a unitary LCS of GSA Γ.

So, any GSA Γ includes exactly M unitary LCSs. Let us denote a unitary LCS (ULCS) corresponding to the state $a_m \in A$ by the symbol a_m ($m = 1, \ldots, M$).

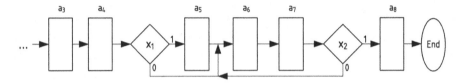

Fig. 3.12 Fragment of GSA Γ_0

Obviously, the fragment of GSA Γ_0 contains 6 different ULCSs. During this step, we do not care about the content of operator vertices. Because of it, there are no microoperations into operator vertices of Γ_0.

If the synthesis of FSM is based on ULCSs, then the state register is used. We discuss corresponding design methods in Chaps. 4 and 5.

Definition 3.2 An elementary LCS (ELCS) of GSA Γ is a finite vector $\alpha_g = \langle a_{g1}, \ldots, a_{gF_g} \rangle$ such that there is unconditional transition $\langle a_{gi}, a_{gF_g} \rangle$ for any pair of adjacent components of the vector α_g.

Let $A(\alpha_g)$ be a set of states which are components of ELCS α_g ($g = 1, \ldots, G1$). Any ULCS has only one input and one output.

Definition 3.3 A state $a_m \in A(\alpha_g)$ is an input of ELCS α_g if the input of operator vertex marked by $a_m \in A$ is connected with output of either start vertex b_0 or conditional vertex or any operator vertex marked by a state $a_s \notin A(\alpha_g)$.

Definition 3.4 A state $a_m \in A(\alpha_g)$ is an input of ELCS α_g if either there is unconditional transition with an input of other ULCS or there is the transition $\langle a_m, a_1 \rangle$ or there are conditional transitions from the state $a_m \in A(\alpha_g)$.

Let us denote an input of ELCS α_g by the symbol I_g, whereas the output of α_g by O_g. Let us consider Fig. 3.12.

The following ULCSs can be constructed: $\alpha_1 = \langle a_3, a_4 \rangle, \alpha_1 = \langle a_5 \rangle, \alpha_3 = \langle a_6, a_7 \rangle$ and $\alpha_4 = \langle a_8 \rangle$. The following inputs and outputs exist for these chains: $I_1 = a_3$, $O_3 = a_4; I_2 = O_2 = a_5; I_3 = a_6, O_3 = a_7; I_4 = O_4 = a_8$.

Definition 3.5 A normal LCS (NLCS) of GSA Γ is a finite vector $\beta_g = \langle a_{g1}, \ldots, a_{gF_g} \rangle$ such that there is unconditional transition $\langle a_{gi}, a_{gi+1} \rangle$ for any pair of adjacent components of β_g.

Definition 3.6 A state $a_m \in A(\beta_g)$ is an input of NLCS β_g if the input of operator vertex marked by $a_m \in A$ is connected with the output of any vertex which is not marked by any state $a_s \in A(\beta_g)$.

It means that the input of the vertex marked by the state a_m should not be connected with outputs of either the start vertex, or conditional vertices or any operator vertex which is marked by state $a_s \notin A(\beta_g)$. Any NLCS $\beta_g(g = 1, \ldots, G2)$ can include up to F_g inputs.

Definition 3.7 An input $a_m \in A(\beta_g)$ is a main input if the input of operator vertex marked by $a_m \in A$ is not connected with the output of any operator vertex of GSA Γ.

Obviously, each NLCS can include only one main input. It corresponds to the first component of the vector β_g ($g = 1, \ldots, G2$).

Definition 3.8 A state $a_m \in A(\beta_g)$ is an output of NLCS β_g if the output of operator vertex marked by a_m is connected with any vertex which is not marked by any state $a_s \in A(B_g)$.

It means that the output of the vertex marked by $a_m \in A(\beta_g)$ is connected with an input of either final vertex or conditional vertex or any operator vertex marked by the state $a_s \notin A(\beta_g)$. Obviously, any NLCS β_g can have exactly one output. Let us denote the input number k of NLCS β_g as I_g^k ($k = 1, \ldots, F_g$; $g = 1, \ldots, G2$). Let us still use the symbol O_g for the output of NLCS β_g.

For the considered fragment of GSA Γ_0, the following chains can be constructed: $\beta_1 = \langle a_3, a_4 \rangle$, $\beta_2 = \langle a_5, a_6, a_7 \rangle$ and $\beta_3 = \langle a_8 \rangle$. They have the following inputs and outputs: $I_1^1 = a_3$, $O_1 = a_4$; $I_2^1 = a_5$, $I_2^2 = a_6$, $O_2 = a_7$; $I_3^1 = O_3 = a_8$. Obviously the inputs I_g^1 are the main inputs of chains β_g ($g = 1, 2, 3$).

It is clear that elementary chains can be viewed as some parts of natural chains. The following equations can be found for the discussed case: $\beta_1 = \alpha_1$; $\beta_2 = \alpha_2 * \alpha_3$; $\beta_3 = \alpha_4$. We use the sign $*$ for the concatenation of chains.

Definition 3.9 An extended LCS (XLCS) of GSA Γ is a finite vector $\gamma_g = \langle a_{g_1}, \ldots, a_{g F_g} \rangle$ such that there either conditional or unconditional transition $\langle a_{gi}, a_{gi+1} \rangle$ for any pair of adjacent components of γ_g.

The main difference of XLCS from NLCS is reduced to the existence of conditional transitions between states inside the same XLCS. For all other types of chains, only unconditional transitions are permitted for components of the same chain.

Definition 3.10 A state $a_m \in A(\gamma_g)$ is an input of XLCS γ_g if the input of operator vertex marked by the state $a_m \in A$ is connected with the output of either start vertex b_0 or any operator vertex marked by a state $a_s \notin A(\gamma_g)$ or marked by the zero output of a conditional vertex.

Definition 3.11 A state $a_m \in A(\gamma_g)$ is an output of XLCS γ_g if the output of operator vertex marked by the state $a_m \in A$ is connected with the input of either final vertex or any operator vertex marked by the state $a_s \notin A(\gamma_g)$ or any conditional vertex of GSA Γ.

Definition 3.12 An input $a_m \in A(\gamma_g)$ is a main input of XLCS γ_g if the input of operator vertex marked by the state $a_m \in A$ is not connected with the output of any operator vertex marked by a state $a_s \notin A(\gamma_g)$.

In the last three definitions, the symbol $A(\gamma_g)$ stands for the set of components of XLCS γ_g ($\gamma = 1, \ldots, G3$). Obviously, the issue of XLCS is the most general issue. Such a chain can include more than one input ($I_g^1, I_g^2, \ldots$) and more than one output ($O_g^1, O_g^1, \ldots$).

In the case of the fragment shown in Fig. 3.12, there is a single XLCS $\gamma_1 = \langle a_3, a_4, \ldots, a_8 \rangle$. It has 4 inputs and 3 outputs: $I_g^1 = a_3$, $I_g^2 = a_5$, $I_g^3 = a_6$, $I_g^4 = a_8$, $O_g^1 = a_4$, $O_g^2 = a_7$ and $O_g^3 = a_8$ ($g = 1$). Obviously, the following expressions are true: $\gamma_1 = \alpha_1 * \alpha_2 * \alpha_3 * \alpha_4$ and $\gamma_1 = \beta_1 * \beta_2 * \beta_3$.

Let us discuss two very important issues:

1. How to construct different LCSs?
2. How they influence structures of corresponding FSMs?

3.4   Structures of LCS-Based FSMs

If a synthesis method for some FSM is based on existing linear chains of states, let us name such an FSM the LCS-based FSM. Let us start from ELCS-based Moore FSMs.

Let it be some GSA Γ_m marked by the states of Moore FSM using the approach from [4]. Let the following sets be constructed for this GSA:

1. The set $C_E = \{\alpha_1, \ldots, \alpha_{G1}\}$ of ELCSs determining a partition of the set of states A.
2. The set I_E of inputs of chains $\alpha_g \in C_1$. This set is determined as

$$I_E = \bigcup_{g=1}^{G1} I_g. \tag{3.24}$$

3. The set O_E of outputs of chains $\alpha_g \in C_1$. This set is determined as

$$O_E = \bigcup_{g=1}^{G1} O_g. \tag{3.25}$$

Let us execute the following state assignment for each pair of adjacent components of chains $\alpha_g \in C_1$:

$$K(a_{gi+1}) = K(a_{gi}) + 1. \tag{3.26}$$

This condition should take place for $g = 1, \ldots, G_1$ and $i = 1, \ldots, F_g - 1$. Let us name the state assignment (3.26) a natural state assignment.

Now, all transitions of FSM (all arcs of corresponding STG) can be derived by two classes:

1. The class T_{in} of transitions between the states inside the chains. These transitions are executed using the rule (3.26). Obviously, it can be done by adding 1 to the content of some counter CT.
2. The class T_{out} of transitions between the outputs and input of chains $\alpha_g \in C_E$. Let us point out that it is possible the conditional transition between an output O_g and an input $I_g (g = 1, \ldots, G1)$. To implement the transitions from T_{out}, it is necessary to load a code of the next state into the counter.

It is necessary to have some additional variable y_0 to control the counter. Let us use the following agreements:

$$y_0 = 1 \rightarrow CT := CT + 1; \qquad (3.27)$$
$$y_0 = 0 \rightarrow CT := \langle \Phi \rangle. \qquad (3.28)$$

The condition (3.28) determines the loading CT from some external block. As always in this book, the counter has informational inputs of the type D.

Let us use the symbol PY_E for determining the ELCS-based Moore FSM. Its structural diagram is shown in Fig. 3.13.

The PY_E Moore FSM operates in the following manner. If there is *Start* = 1, then the code of initial state $a_1 \in A$ is loaded into the CT. Every pulse of CT permits generating the collections of microoperations $Y_t \subseteq Y$ by the block of microoperations BMO. Let some code $K(a_m)$ be in the CT in some instant $t (t = 0, 1, \ldots)$. If $a_m \neq O_g (g = 1, \ldots, G1)$, then the variable $y_0 = 1$ is generated by the BMO. It causes incrementing the counter CT (see (3.27)). If $a_m = O_g (g = 1, \ldots, G1)$ then the CT is loaded by the block of input memory functions BIMF. If the next state $a_s = a_1$, then the content of CT is not changed till the next pulse *Start* arrives.

So, in PY_E Moore FSM the BIMF generates functions (2.9), whereas the BMO generates functions (2.16) and the additional variable

$$y_0 = y_0(T). \qquad (3.29)$$

The set C_E should correspond to the following conditions:

Fig. 3.13 Structural diagram of PY_E Moore FSM

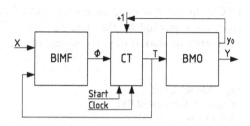

$$A(\alpha_g) \neq \emptyset \ (g = 1, \ldots, G1); \tag{3.30}$$

$$\bigcup_{g=1}^{G1} A(\alpha_g) = A \setminus \{a_1\}; \tag{3.31}$$

$$A(\alpha_i) \cap A(\alpha_j) \neq \emptyset \ (i \neq j, i, j \in \{1, \ldots, G1\}; \tag{3.32}$$

$$G1 \rightarrow \min. \tag{3.33}$$

These conditions mean the following:

1. Any ELCS $\alpha_g \in C_E$ includes at least a single state $a_m \in A$ (condition (3.30)).
2. There are no states $a_m \in A$ which are not included in some chain $\alpha_g \in C_E$. The only exception is the initial state $a_1 \in A$ (conditions (3.31)).
3. Different states $a_m \in A$ are included into different chains $\alpha_g \in C_1$ (condition (3.32)). It allows maximizing the number of rows of structure table having no input memory functions.
4. The set C_E includes the minimal possible number of chains (condition (3.33)). It has the same effect as the condition (3.32).

So, the set C_E is a partition of the set A by the chains $\alpha_g \ (g = 1, \ldots, G1)$ with minimal amount of classes $A(\alpha_g) \subseteq A$. Obviously, the following relation takes place:

$$1 \leq G1 \leq M - 1. \tag{3.34}$$

If there is $G1 = 1$, then each class $A(\alpha_g)$ includes only a single state. In this case, there is no need in a counter and the model of PY Moore FSM should be used. If there is $G1 = M - 1$, then the corresponding GSA Γ_m does not include conditional vertices. It leads to degenerated PY_E Moore FSM (Fig. 3.14).

Obviously, the design of FSM (Fig. 3.14) is trivial. We do not discuss it in this book.

The set C_E corresponding to (3.30)–(3.33) can be constructed in two steps:

1. Finding the set of inputs I_E for a given GSA.
2. Constructing an ELCS for each element of I_E.

The set I_E is constructed using Definition 3.3. There are $G1 = 7$ inputs of ELCSs in the discussed case (Fig. 3.15). Let the symbol $b(a_m)$ means an operator vertex marked by the state $a_m \in A$. The input of $b(a_2)$ is connected with the output of start vertex; the inputs of vertices $b(a_4)$, $b(a_5)$, $b(a_6)$, $b(a_8)$, $b(a_9)$

Fig. 3.14 Structural diagram of degenerated PY_E Moore FSM

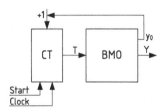

Fig. 3.15 Initial
graph-scheme of algorithm
Γ_4

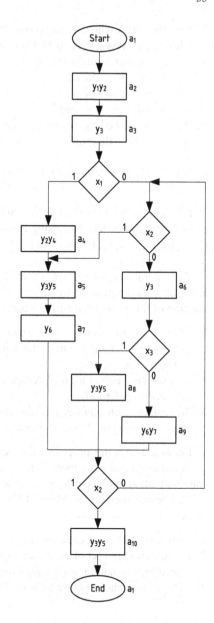

and $b(a_{10})$ are connected with outputs of conditional vertices. So, there is the set
$I_E = \{a_2, a_4, a_5, a_6, a_8, a_9, a_{10}\}$; it means that $G1 = 7$.

The following procedure is proposed for constructing the partition C_E:

1. Put $g = 1$.
2. Take the state with the smallest number m from the set I_E. Exclude this state from
the set I_E. Let us name this state as a base state of ELCS $\alpha_g \in C_E$.

3. Find the vertex whose input is connected with the output of operator vertex $b(a_m)$ where a_m is a base state of ELCS $\alpha_g \in C_E$. Let us denote this vertex as b_{next}.
4. If b_{next} is either a conditional or final vertex, then the construction of ELCS α_g is terminated. Go to the step 7.
5. If b_{next} is marked by a state $a_s \in I_E$, then the construction of the chain α_g is terminated. Go to the step 7.
6. If b_{next} is marked by a state $a_s \notin I_E$, then this state is included into the chain $\alpha_g \in C_E$. Now this state is considered as a base state a_m. Go to the step 3.
7. $g = g + 1$.
8. If $g < G1$, then go to step 9 else go to the step 2.
9. End.

Let us point out that initial set I_E is used during the steps 5 and 6. Let us name this procedure as procedure P1. Using P1 for discussed case leads to the following set $C_E = \{\alpha_1, \ldots, \alpha_7\}$. This set includes the following chains: $\alpha_1 = \langle a_2, a_3 \rangle, \alpha_2 = \langle a_4 \rangle$, $\alpha_3 = \langle a_5, a_7 \rangle, \alpha_4 = \langle a_6 \rangle, \alpha_5 = \langle a_8 \rangle, \alpha_6 = \langle a_9 \rangle$, and $\alpha_7 = \langle a_{10} \rangle$.

The natural state assignment should be executed using R-dimensional codes. The value of R is determined by (2.15). The following procedure P2 is proposed for solution of this problem for ELCS-based Moore FSMs:

1. To construct the vector $\alpha = \alpha_1 * \alpha_2 * \ldots * \alpha_{G1}$, where $*$ is a sign of concatenation. Let us point out that the first ELCS in α should include a state a_m such that there is a transition $\langle a_1, a_m \rangle$.
2. To execute a numeration of components of the vector α using the consecutive integers from 1 to M.
3. To replace each number i of the component a_m of the vector α by its binary equivalent having R bits. The final codes are treated as the codes $K(a_m)$.

Let us illustrate this procedure for the discussed example. It is shown in Table 3.4. In the discussed case, there is $R = 4$. Obviously, the initial state $a_1 \in A$ has the code with all zeros. The process of state assignment is obvious from Table 3.4.

The design method for Moore FSM with PY_E structure includes the following steps:

1. Constructing the set of states A.
2. Constructing the set of elementary LCSs C_E.
3. Executing the natural state assignment.
4. Constructing the structure table of PY_E Moore FSM.

Table 3.4 Natural state assignment for Moore FSM $PY_E(\Gamma_4)$

Steps	α_g	α_1		α_2	α_3		α_4	α_5	α_6	α_7
1	a_m	a_2	a_3	a_4	a_5	a_7	a_6	a_8	a_9	a_{10}
2	i	1	2	3	4	5	6	7	8	9
3	$K(a_m)$	0001	0010	0011	0100	0101	0110	0111	1000	1001

5. Constructing the system of input memory functions.
6. Constructing the table of microoperations.
7. Implementing the FSM logic circuit.

Let us discuss an example of design for the Moore FSM $PY_E(\Gamma_4)$. The sets A and C_E are obtained before, the state codes can be found from Table 3.4.

To construct the structure table, let us construct the system of formulae of transitions [9] for states $a_1 \in A$ and $a_m \in O_E$, where $O_E \subseteq A$ is a set of outputs of ELCSs $\alpha \in C_E$.

In the discussed case, there is the set $O_E = \{a_3, a_4, a_6, \ldots, a_{10}\}$. The following system of formulae of transitions (SFT) could be found from GSA Γ_4:

$$
\begin{aligned}
a_1 &\rightarrow a_2 \\
a_3 &\rightarrow x_1 a_4 \vee \bar{x}_1 x_2 a_5 \vee \bar{x}_1 \bar{x}_2 a_6; \\
a_4 &\rightarrow a_5; \\
a_6 &\rightarrow x_3 a_8 \vee \bar{x}_3 a_9; \\
a_7 &\rightarrow x_2 a_{10} \vee \bar{x}_2 a_6; \\
a_8 &\rightarrow x_2 a_{10} \vee \bar{x}_2 a_6; \\
a_9 &\rightarrow x_2 a_{10} \vee \bar{x}_2 a_6; \\
a_{10} &\rightarrow a_1.
\end{aligned}
\qquad (3.35)
$$

The structure table of Moore FSM $PY_E(\Gamma_4)$ includes $H_E(\Gamma_4) = 12$ rows (Table 3.5). Let us point out that the transition from a_{10} into a_1 is executed by the pulse Clock. So, the structure table does not include the corresponding row. Also, the transition from a_1 into a_2 is executed using y_0. So, this row is not included in Table 3.5.

Table 3.5 Structure table of Moore FSM $PY_E(\Gamma_4)$

a_m	$K(a_m)$	a_s	$K(a_s)$	X_h	Φ_h	h
a_3	0010	a_4	0011	x_1	$D_3 D_4$	1
		a_5	0100	$\bar{x}_1 x_2$	D_2	2
		a_6	0110	$\bar{x}_1 \bar{x}_2$	$D_2 D_3$	3
a_4	0010	a_5	0100	1	D_2	4
a_6	0110	a_8	0111	x_3	$D_2 D_3 D_4$	5
		a_9	1000	$\bar{x}_3$	D_1	6
a_7	0101	a_{10}	1001	x_2	$D_1 D_4$	7
		a_6	0110	$\bar{x}_2$	$D_2 D_3$	8
a_8	0111	a_{10}	1001	x_2	$D_1 D_4$	9
		a_6	0110	$\bar{x}_2$	$D_2 D_3$	10
a_9	1000	a_{10}	1001	x_2	$D_1 D_4$	11
		a_6	0110	$\bar{x}_2$	$D_2 D_3$	12

Table 3.6 Table of microoperations of Moore FSM $PY_E(\Gamma_4)$

$K(a_m)\, T_1 T_2 T_3 T_4$	y_0	$Y_m\ y_1 y_2 y_3 y_4 y_5 y_6 y_7$	m
0000	1	000 0000	1
0001	1	110 0000	2
0010	0	001 0000	3
0011	0	010 1000	4
0100	1	001 0100	5
0101	0	000 0010	6
0110	0	001 0000	7
0111	0	001 1000	8
1000	0	000 0011	9
1001	0	001 1000	10

The system Φ is constructed in the traditional way. For example, the following formula can be derived for the function D_1 (after minimizing the initial SOP):

$$D_1 = F_6 \vee [F_7 \vee F_9] \vee F_{11} = \bar{T}_1 T_2 T_3 \bar{T}_4 \bar{x}_3 \vee \bar{T}_1 T_2 T_4 x_2 \vee T_1 \bar{T}_2 \bar{T}_3 \bar{T}_4 x_2. \quad (3.36)$$

Let us point out that (3.36) can be minimized using the "don't care" input assignments from 1010 till 1111. Taking them into account, the following minimized expression can be obtained for the function D_1:

$$D_1 = T_2 T_3 \bar{T}_4 \bar{x}_3 \vee T_2 T_4 x_2 \vee T_1 \bar{T}_4 x_2. \quad (3.37)$$

The expressions similar to (3.37) can be obtained for any function $D_r \in \Phi$.

The table of microoperations can be constructed using a marked GSA. It includes the following columns: $K(a_m)$, y_0, Y_m, m. In the discussed case, this table includes 16 rows. Only 10 of them include some useful information. Namely these rows are shown in Table 3.6.

The column y_0 is filled in the following manner. If $a_m \notin O_E$, then $y_0 = 1$ in the corresponding row of the table. Because of unconditional transition $\langle a_1, a_2 \rangle$, there is $y_0 = 1$ in the first row of the table.

Now, the obtained input memory functions $D_r \in \Phi$ should be implemented using LUTs. The table of microoperations determines the content of EMBs implementing the block BMO. We do not discuss this step.

Now, let us discuss the design of NLCS-based Moore FSMs. Let the following sets be obtained for some marked GSA Γ_i:

1. The set $C_N = \{\beta_1, \ldots, \beta_{G2}\}$ of NLCS determining a partition of the set of states A.
2. The set I_N of inputs of chains $\beta_g \in C_N$. This set is determined as

$$I_N = \bigcup_{g=1}^{G2} I(\beta_g). \qquad (3.38)$$

In (3.38), the symbol $I(\beta_g) \subseteq A(\beta_g)$ is a set of inputs of an NLCS $\beta_g \in C_N$.
3. The set O_N of outputs of chains $\beta_g \in C_2$. This set is determined as

$$O_N = \bigcup_{g=1}^{G2} O_g. \qquad (3.39)$$

Let the natural state assignment be executed for states $a_m \in A$. In this case, the transitions among the states $a_m \in A$ could be derived by the classes T_{in} and T_{out} (as it is done for PY_E Moore FSM). Because of the state assignment (3.26), the variable y_0 is necessary, too. Its functions are determined by (3.27)–(3.28).

Let us use the symbol PY_N for determining the NLCS-based Moore FSM. Its structural diagram is same as the one for PY_E Moore FSM. Obviously, the principles of operations are the same for these both models. To show an advantage of PY_E Moore FSM, we should discuss the principle of code sharing [8]. It will be done a bit later.

The set C_N should correspond to the following conditions:

$$A(B_g) \notin \varnothing \ (g = 1, \dots, G2); \qquad (3.40)$$

$$\bigcup_{g1}^{G2} A(\beta_g) = A \setminus \{a_1\}; \qquad (3.41)$$

$$A(\beta_i) \cap A(\beta_j) = \varnothing \ (i \neq j; i, j \in \{1, \dots, G2\}); \qquad (3.42)$$

$$G2 \rightarrow \min. \qquad (3.43)$$

The meaning of conditions (3.40)–(3.43) is the same as for conditions (3.30)–(3.33), respectively. Of course, the former represent the properties of NLCSs $\beta_g \in C_N$.

The set C_N corresponding to (3.40)–(3.43) can be constructed in two steps:

1. Finding the set of main inputs IM_N for a given GSA Γ.
2. Constructing an NLCS for each element of IM_N.

Let us discuss the execution of these steps for the GSA Γ_4. The set IM_N is constructed using Definition 3.7. The input of operator vertex $b(a_2)$ is connected with the output of the start vertex. The inputs of vertices $b(a_4), b(a_6), b(a_8), b(a_9)$ and $b(a_{10})$ are connected only with outputs of conditional vertices. It means that there is the set $IM_N = \{a_2, a_4, a_8, a_9, a_{10}\}$. So, there is $G2 = 6$.

The following procedure P3 is proposed for constructing the partition C_N:

1. Put g=1.
2. Take the state with the smallest number of m from the set IM_N. Exclude this state from the set IM_N. Let us name this state a base state for NLCS $\beta_g \in C_N$.

3. Find the vertex b_{next} (as for P1).
4. If b_{next} is either conditional or final vertex, then construction of the chain β_g is terminated. Go to the step 6.
5. If b_{next} is not marked by a state a_s already included into some chain $\beta_g \in C_N$, then this state is included into the chain β_g. Now, this state is considered as the base state a_m. Go to step 3.
6. $g = g + 1$.
7. If $IM_N \neq \emptyset$, then go to the step 2.
8. End.

Let us apply the procedure P3 to GSA Γ_4. It gives the set $C_N = \{\beta_1, \ldots, \beta_6\}$ where $\beta_1 = \langle a_2, a_3 \rangle$, $\beta_2 = \langle a_4, a_5, a_7 \rangle$, $\beta_3 = \langle a_6 \rangle$, $\beta_4 = \langle a_8 \rangle$, and $\beta_5 = \langle a_{10} \rangle$. In the case of PY_N Moore FSM, the procedure P2 is used for executing the natural state assignment. In this particular case, the outcomes of P2 coincide for PY_{E4} and PY_{N4}. So, the natural states codes for Moore FSM PY_{N4} are shown in the last row of Table 3.4.

The design methods for PY_E and PY_N FSMs are the same. But the execution of the steps 2 and 4 is different. During the step 2, the set C_N is constructed. During the step 4, the SFT is constructed for the set $O_N = \{a_3, a_6, a_7, a_8, a_9, a_{10}\}$. In this case, the structure table of PY_{N4} includes only $H_N(\Gamma_4) = 11$ rows. The only difference in the table of microoperations is reduced to existence of 1 in the column y_0 for $m = 4$. We do not show these tables in that chapter.

Now, let us discuss the design of XLCS-based Moore FSMs. Let the following sets be obtained for some marked GSA Γ_m:

1. The set $C_X = \{\gamma_1, \ldots, \gamma_{G3}\}$ of XLCS-based Moore FSMs.
2. The set I_X of inputs of chains $\gamma_g \in C_X$. This set is determined as

$$I_X = \bigcup_{g=1}^{G3} I(\gamma_g). \tag{3.44}$$

In (3.44), the symbol $I(\gamma_g)$ denotes a set of inputs of an XLCS $\gamma_g \in C_X$.
3. The set O_X of outputs of chains $\gamma_g \in C_3$. This set is determined by the following expression:

$$O_X = \bigcup_{g=1}^{G3} O(\gamma_g). \tag{3.45}$$

In (3.45), the symbol $O(\gamma_g)$ stands for the set of outputs of an XLCS $\gamma_g \in C_3$.

Let the natural state assignment be executed for states $a_m \in A$. In this case, the transitions among the states could be derived by the classes T_{in} and T_{out}. Due to the state assignment (3.26), the variable y_0 is necessary.

Fig. 3.16 Structural
diagram of PY_X Moore FSM

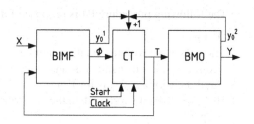

Let us use the symbol PY_X for determining the XLCS-based Moore FSM. Accord-
ing with Definition 3.9, an XLCS can have more than one output and condition
(3.26) should take place for conditional transitions, too. It means that the variable
y_0 can be represented as

$$y_0 = y_0^1 \vee y_0^2. \tag{3.46}$$

The variable $y_0^1 (y_0^2)$ is generated by BIMF(BMO). It determines the following structure of PY_X Moore FSM (Fig. 3.16).

The main difference of PY_X FSM from PY_E and PY_N FSMs is existence of the
following functions:

$$y_0^1 = f_1(X, T); \tag{3.47}$$
$$y_0^2 = f_2(T). \tag{3.48}$$

We do not discuss the mode of operation of PY_X FSM. It is rather obvious.

The set C_X should correspond to the following conditions:

$$A(\gamma_g) \neq \emptyset \, (g = 1, \ldots, G3); \tag{3.49}$$

$$\bigcup_{g=1}^{G3} A(\gamma_g) = A \setminus \{a_1\}; \tag{3.50}$$

$$A(\gamma_i) \cap A(\gamma_j) = \emptyset \, (i \neq j; i, j \in \{1, \ldots, G3\}); \tag{3.51}$$

$$G3 \rightarrow \min. \tag{3.52}$$

The set C_X corresponding to (3.49)–(3.52) can be constructed in two steps:

1. Finding the set of main inputs IM_X for a given GSA Γ.
2. Constructing an XLCS for each input $I_g^1 \in IM_X$.

Let us discuss the execution of these steps for the GSA Γ_4. The set IM_X is
constructed using Definition 3.12. The vertex $b(a_2)$ is connected with outputs of
conditional variables marked by zero. So, there is the set $IM_X = \{a_2, a_6, a_9\}$ and
$G3 = 3$.

The following procedure P4 is proposed for constructing the partition C_X.

1. Put g=1.
2. Take the state with the smallest number m from the set IM_X. Exclude this state from IM_X. Let us name this state a base state for XLCS $\gamma_g \in C_X$.
3. Find the vertex b_{next} (as for P1).
4. If b_{next} is the end vertex, then the constructing the chain γ_g is terminated. Go to step 10.
5. If b_{next} is an operator vertex $b(a_s)$ such that the state a_s is already included in some other chain, then the constructing the chain γ_g is terminated. Go to step 10.
6. If b_{next} is a conditional vertex, then find an operator vertex $b(a_s)$ connected with marked by 1 outputs of conditional vertices marking a path of GSA [4] starting from the vertex b_{next}.
7. If there is no such an operator vertex (the path is finished by the end vertex), then the constructing the chain γ_g is terminated. Go to step 10.
8. If the state a_s (from step 6) is already included into some other chain, then the constructing the chain γ_g is terminated. Go to step 10.
9. If the state a_s (from either step 5 or step 8) is not included in some other chain, then this state is included into the chain γ_g. Now, this state is considered as the base state a_m. Go to step 4.
10. $g = g + 1$.
11. If $IM_X \neq \emptyset$, then go to step 2.
12. End.

Application of P4 to GSA Γ_4 produces the set of XLCSs $C_X = \{\gamma_1, \gamma_2, \gamma_3\}$. It includes the following chains $\gamma_1 = \langle a_2, a_3, a_4, a_5, a_7, a_{10}\rangle, \gamma_2 = \langle a_2, a_8\rangle, \gamma_3 = \langle a_9\rangle$.

As in the previous cases, the natural state assignment is executed by the procedure P2. In the case of $PY_X(\Gamma_4)$, it gives the following outcome (Table 3.7).

The design method for PY_X More FSM includes the following steps:

1. Constructing the set of states A.
2. Constructing the set of extended LCSs C_X.
3. Executing the natural state assignment.
4. Constructing the structure table of PY_X Moore FSM.
5. Constructing the system of input memory functions and the function y_0^1.
6. Constructing the table of microoperations.
7. Implementing the FSM logic circuit.

We discuss this method in details in Chap. 8.

Table 3.7 Natural state assignment for Moore FSM $PY_X(\Gamma_4)$

Step	γ_g	γ_1						γ_2		γ_3
1	a_m	a_2	a_3	a_4	a_5	a_7	a_{10}	a_6	a_8	a_9
2	i	1	2	3	4	5	6	7	8	9
3	$K(a_m)$	0001	0010	0011	0100	0101	0110	0111	1000	1001

3.5 Principles of Hardware Reduction for LCS-Based Finite State Machines

To generalize the discussed here principles, let us define the object of FSM. It can be:

1. state $a_m \in A$;
2. LCS;
3. class of states;
4. class of LCSs;
5. collection of microoperations.

To reduce the number of LUTs in the circuit of BIMF, it is necessary to diminish the numbers of arguments and terms in the system of input memory functions [9]. One of the possible approaches for solution of this problem is a code sharing [8]. Let us discuss this principle for cases of NLCS and ELCS-based FSMs.

Let some GSA Γ include $G2$ NLCSs $\beta_g \in C_N$. Let us encode each chain $\beta_g \in C_N$ by a binary code $K(\beta_g)$ having R_{G2} bits:

$$R_{G2} = \lceil \log_2 G2 \rceil. \tag{3.53}$$

States $a_m \in A$ are distributed among the sets $A(\beta_g)$. Let it be $M_g^2 = |A(\beta_g)|$. Let us find the maximal amount of states $M_{G2} = \max(M_1^2, \ldots, M_{G2}^2)$. Let us encode each state $a_m \in A$ by a binary code $C(a_m)$ having R_{C2} bits:

$$R_{C2} = \lceil \log_2 M_{G2} \rceil. \tag{3.54}$$

Let us point out that different states $a_m \in A$ can have the same codes $C(a_m)$. But any chain $\beta_g \in C_2$ should include only the states with different codes $C(a_m)$. Let the following rule be used for the codes $C(a_m)$:

$$C(a_{gi+1}) = C(a_{gi}) + 1 \quad (g = 1, \ldots, G2; i = 1, \ldots, F_g - 1). \tag{3.55}$$

The condition (3.55) means that the natural state assignment is executed for the states $a_m \in A(\beta_g)$ $(g = 1, \ldots, G2)$. Let us use the variables $\tau_r \in \tau$ for encoding the chains, where $|\tau| = R_{G2}$. Let us use the variables $T_r \in T$ for state assignment, where $|T| = R_{C2}$.

The approach allows to represent any code $K(a_m)$ as the following concatenation:

$$K(a_m) = K(\beta_g) * C(a_m). \tag{3.56}$$

In (3.56), the sign * is used for the operation of concatenation.

Such a representation of the code of any object is named the **code sharing**. The formula (3.56) determines the structure diagram of PY_{NC} Moore FSM (Fig. 3.17).

Fig. 3.17 Structural diagram
of PY$_{NC}$ Moore FSM

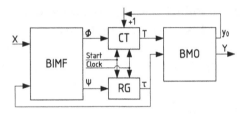

In PY$_{NC}$ FSM, the block BIMF implements input memory functions loaded into
the counter CT(Φ) and register RG(Ψ):

$$\Phi = \Phi(\tau, X); \tag{3.57}$$

$$\Psi = \Psi(\tau, X). \tag{3.58}$$

The block BMO implements the functions

$$Y = Y(\tau, X); \tag{3.59}$$

$$y_0 = y_0(\tau, X). \tag{3.60}$$

Let the following condition take place:

$$R_{G2} < R. \tag{3.61}$$

In this case, the systems Φ and Ψ of PY$_{NC}$ FSM include few arguments than the
system Φ for PY Moore FSM. The best case for applying the code sharing for
NLCS-based FSM is determined by the following relation:

$$R_{G2} + R_{C2} = R. \tag{3.62}$$

Let us point out that modern EMBs are very powerful. It means that only a single
EMB is enough for implementing the systems (3.59)–(3.60) even if condition (3.62)
is violated.

This very principle can be used for ELCS-based Moore FSMs. Let us replace
Eqs. (3.53)–(3.56) by the following equations:

$$R_{G1} = \lceil \log_2 G1 \rceil; \tag{3.63}$$

$$R_{C1} = \lceil \log_2 M_{G1} \rceil; \tag{3.64}$$

$$C(a_{gi+1}) = C(a_{gi}) + 1 \quad (g = 1, \ldots, G1; i = 1, \ldots F_g - 1); \tag{3.65}$$

$$K(a_m) = K(\alpha_g) * C(a_m). \tag{3.66}$$

The meaning is obvious for each element of formulae (3.63)–(3.66). Let us use
the variables τ_r for encoding of the chains $\alpha_g \in C_E$. Let us use the variables T_r for

Fig. 3.18 Structural diagram of PY_{EC} Moore FSM

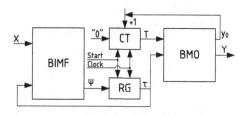

encoding of the components $a_m \in A(\alpha_g)$. The set τ includes G_{G1} elements, the set T includes R_{C1} elements.

In the case of elementary LCSs, each of them has only a single input. Let us start the natural state assignment for each ELCS $\alpha_g \in C_1$ from the code $C(a_m)$ having all zeros. It means that it is enough to connect all informational inputs of D flip-flops of CT with values of logic zeros. In this case, there is no need in functions Φ. Now, the structural diagram of PY_{EC} Moore FSM can be obtained (Fig. 3.18).

The block BIMF of PY_{EC} FSM implements only R_{C1} functions. Because of $R_{C2} \leq R$, it is the minimal possible number of functions.

The second issue of possible hardware reduction is the existence of **pseudoequivalent objects**. Every pair of objects with equivalent transitions determines two pseudoequivalent objects. Obviously, this relation is reflexive (any object is pseudoequivalent to itself), symmetric (if some object A is pseudoequivalent to some object B, then the object B is pseudoequivalent to the object A) and transitive (if some object A is pseudoequivalent to some object B which is pseudoequivalent to some object C, then the object A is pseudoequivalent to the object C). As it is known [4], such kind of relations determine some partition of the set of objects.

The following partitions can be found for GSA γ_4. The partition $\Pi_A = \{B_1, \ldots, B_8\}$ includes the classes of PES. There are the following classes: $B_1 = \{a_1\}$, $B_2 = \{a_2\}$, $B_3 = \{a_3\}$, $B_4 = \{a_4\}$, $B_5 = \{a_5\}$, $B_6 = \{a_6\}$, $B_7 = \{a_7, a_8, a_9\}$ and $B_8 = \{a_{10}\}$. The partition $\Pi_{CE} = \{B_1, \ldots, B_5\}$ includes the classes of pseudoequivalent ELCSs $\alpha_g \in C_E$. There are the following classes: $B_1 = \{\alpha_1\}$, $B_2 = \{\alpha_2\}$, $B_3 = \{\alpha_3, \alpha_5, \alpha_6\}$, $B_4 = \{\alpha_4\}$, $B_5 = \{\alpha_7\}$. The partition $\Pi_{CN} = \{B_1, \ldots, B_4\}$ includes the classes of pseudoequivalent NLCS $\beta_g \in C_2$. There are the following classes: $B_1 = \{\beta_1\}$, $B_2 = \{\beta_1, \beta_4, \beta_5\}$, $B_3 = \{\beta_3\}$, $B_4 = \{\beta_6\}$. There is no partition Π_{CX} having at last a single class with more than 1 XLCS $\gamma_g \in C_X$. It is explained by existing more than one output in the class XLCSs γ_1 and γ_2.

Two approaches are used for taking into account the existence of pseudoequivalent objects. The first is the optimal encoding of objects. It assumes such an encoding when a class is represented by minimum possible amount of generalized intervals of Boolean space. It is the optimal state assignment in the case of PY Moore FSM.

In the case of $P_0Y(4)$, the optimal state codes are shown in Fig. 3.19.

The following codes of classes $B_i \in \Pi_A$ can be derived from the Karnaugh map (Fig. 3.19): $K(B_1) = *000$, $K(B_2) = 0010$, $K(B_3) = 011*$, $K(B_4) = *1*1$, $K(B_5) = *100$, $K(B_6) = 101*$, $K(B_7) = *0*1$ and $K(B_8) = 1***$.

T_1 \ $T_2T_3T_4$	000	001	011	010	110	111	101	100
0	a_1	a_7	a_8	a_2	a_3	*	a_4	a_5
1	*	a_9	*	a_6	a_{10}	*	*	*

Fig. 3.19 Optimal state codes of Moore FSM $P_0 Y_{(4)}$

The second approach is based on transformation of object into the classes of pseudoequivalent objects. It assumes using additional variables for encoding of the classes, as well as some special code transformer. As example of this approach, the $P_C Y$ Moore FSM can be taken (Fig. 3.10).

We discuss peculiarities of all these issues in the next chapters of this book. Let us use the following denotations: EFSM is an FSM based on ELCS; NFSM is an FSM based on NLCS; XFSM is an FSM based on XLCS.

References

1. Agrawala, A.K., Rauscher, T.G.: Foundations of Microprogramming. Academic Pres Inc, New York (1976)
2. Amann, R., Baitinger, U.G.: Optimal state chains and state codes in finite state machines. Trans. Comput. Aided Des. Integ. Cir. Sys. **2**(9):153–170 (2006)
3. Amann, R., Eschermann, B., Baitinger, U.G.: PLA based finite state machines using johnson counters as state memories. In: Proceedings of the 1988 IEEE International Conference on Computer Design: VLSI in Computers and Processors, 1988. ICCD'88, pp. 267–270 (1988)
4. Baranov, S.: Logic and System Design of Digital Systems. TUT Press, Tallinn (2008)
5. Baranov, S., Levin, I., Keren, O., Karpovsky, M.: Designing fault tolerant fsm by nano-PLA. In: IEEE International On-Line Testing Symposium (IOLTS 2009), pp. 229–234. Sesimbra-Lisbon, Portugal (2009)
6. Baranov, S., Sklyarov, V.: Digital Devices with Programmable LSI with Matrix Structure. Radio and Communications, Moscow (1986). (in Russian)
7. Barkalov, A.: Microprogram control unit as a composition of automata with programmable and hardwired logic. Autom. Control Comput. Sci. **17**(4), 36–41 (1983)
8. Barkalov, A., Titarenko, L.: Logic synthesis for compositional microprogram control units. Lecture Notes in Electrical Engineering, vol. 22. Springer-Verlag, Berlin (2009)
9. Barkalov, A., Titarenko, L.: Logic Synthesis for FSM-based Control Units. Springer, Berlin (2009)
10. Barkalov, A., Węgrzyn, M.: Design of Control Units with Programmable Logic. University of Zielona Góra Press (2006)
11. Chu, Y.: Computer Organization and Microprogramming. Prentice-Hall Inc, Englewood Cliffs, NJ USA (1972)
12. Czerwiński, R., Kania, D.: Synthesis of finite state machines for CPLDs. Int. J. Appl. Math. Comput. Sci. **19**(4), 647–659 (2009)
13. Dehon, A.: Nanowire-based programmable architectures. ACM J. Emerg. Technol. Comput. Syst. (JETC), **1**(2):109–162 (2005)

14. DeHon, A., Wilson, M.J.: Nanowire-based sublithographic programmable logic arrays. In: Proceedings of the 2004 ACM/SIGDA 12th International Symposium on Field Programmable Gate Arrays, pp. 123–132. ACM (2004)
15. Dehon, A.: Nanowire-based programmable architectures. ACM J. Emerg. Technol. Comput. Syst. (JETC) 1(2), 109–162 (2005)
16. Denisenko, E., Palagin, A., Rokitskii, A.: Design of an addressing system for microprgrammed control devices. Control Syst. Mach. 4, 33–36 (1983). (in Russian)
17. Devjatkov, V.: Methods of implementing finite states machines with shift registers. Nauka (1974). (in Russian)
18. Garcia-Vargas, I., Senhadji-Navarro, R., Jimenez-Moreno, G., Civit-Balcells, A., Guerra-Gutierrez, P.: ROM-based finite state machine implementation in low cost FPGAs. In 2007 IEEE International Symposium on Industrial Electronics, pp. 2342–2347. IEEE (2007)
19. Golson, S.: One-hot state machine design for FPGAs. In: Proceedings of the 3rd Annual PLD Design Conference & Exhibit, vol. 1 (1993)
20. Gorman, K.: The programmable logic array: a new approach to microprogramming. Electron. Des. News 22, 68–75 (1973)
21. Grout, I.: Digital Systems Design with FPGAs and CPLDs. Newnes (2011)
22. Hemel, A.: The PLA: a different kind of ROM. Electron. Des. 24(1), 28–47 (1976)
23. Husson, S.: Microprogramming: Principles and Practices. Prentice Hall (1970)
24. Kirpichnikov, V.M., Sklyarov, V.A.: Synthesis of microprogramming automata described by graph-schemes with small number of conditional nodes. control systems and machines. Naukova Dumka 1, 77–83 (1978). (in Russian)
25. Krishnamoorthy, S., Tessier, R.: Technology mapping algorithms for hybrid FPGAs containing lookup tables and PLAs. IEEE Trans. Comput. Aided Des. Integ. Circuits Syst. 22(5), 545–559 (2003)
26. Lagoviev, B., Lapina, T.: Algorithm of synthesis of automata with a complex register as a memory. Mehanization Autom. Control 2, 45–49 (1976). (in Russian)
27. Maxfield, C.: The Design Warrior's Guide to FPGAs: Devices, Tools and Flows. Elsevier (2004)
28. McCluskey, E.J.: Logic Design Principles: with Emphasis on Testable Semicustom Circuits. Prentice-Hall international editions, Prentice Hall (1986)
29. De Micheli, G.: Synthesis and Optimization of Digital Circuits. McGraw-Hill, New York (1994)
30. Papachristou, C.: Hardware microcontrol schemes using PLAs. In: Proceedings of the 14th Microprogramming Workshop, pp. 3–16 (1982)
31. Papachristou, C.: A PLA microcontroller using horizontal firmware. Microprocess. Microprogr. 14(3), 223–230 (1984)
32. Shrestha, A.M.S., Takaoka, A., Satoshi, T.: On two problems of nano-PLA design. IEICE Trans. Inf. Syst. 94(1), 35–41 (2011)
33. Singh, S., Singh, R.K., Bhatia, M.P.S.: Performance evaluation of hybrid reconfigurable computing architecture over symmetrical FPGA. In: Int. J. Embed. Syst. Appl. 2(3), 107–116 (2012)
34. Sklyarov, V.: Synthesis and implementation of RAM-based finite state machines in FPGAs. In: International Workshop on Field Programmable Logic and Applications, pp. 718–727. Springer (2000)
35. Spivacks, S.: Implementation of microprogrammed automata with counters. Autom. Telemech. 12, 133–139 (1970). (in Russian)
36. Villa, T., Sngiovanni-Vincentelli, A.: Nova: state assignment of finite state machines for optimal two-level logic implementation. IEEE Trans. Comput. Aided Des. Integ. Circuits Syst. 9(9), 905–924 (1990)
37. Xilinx Corp. Web Site (2015). http://www.xilinx.com

Chapter 4
Hardware Reduction for Moore UFSMs

Abstract The Chapter is devoted to the problems of hardware reducing for FPGA-based logic circuits of Moore FSMs. The design methods are proposed based on using more than one source of codes of classes of pseudoequivalent states (PES). Two structural diagrams and design methods are proposed for Moore FSM based on transformation of objects. The first method is based on transformation the unitary codes of microoperations into the codes of PES. The second approach is connected with transformation of the codes of collections of microoperations into the codes of PES. The last part of the Chapter is devoted to the replacement of logical conditions.

4.1 Using Two and Three Sources of Class Codes

As it is pointed out before, the embedded memory blocks of FPGAs are reconfigurable [7]. They have the constant size (V_0) but the number of both cells (V) and outputs (t_F) could be different. The following relation takes place:

$$V = \left\lceil \frac{V_0}{t_F} \right\rceil . \tag{4.1}$$

Some methods are discussed in [5] targeting hardware reduction in CPLD-based Moore FSMs. The methods are based on using up to three sources of codes of classes $B_i \in \Pi_A$. These sources are the state register RG, the block of code transformer BCT and the block of microoperations BMO. The following seven situations are possible (Table 4.1).

If some block produces the class code, then the corresponding cell of Table 4.1 contains the symbol "•". There are more than one source for models $P_C Y_1$–$P_C Y_3$. Using these models is based on the wide fan-in of PAL-based macrocells of CPLD [6]. The acronym PAL stands for programmable array logic. But LUTs have a very limited fan-in (up to 8). It means that the PAL-based models cannot be directly used for the hardware reduction in LUT-based FSMs. Let us discuss these approaches.

Let us construct the partition $\Pi_A = \{B_1, \ldots, B_I\}$ for the set A. Each class $B_i \in \Pi_A$ includes pseudoequivalent states. Let us construct a system of functions

© Springer International Publishing AG 2018
A. Barkalov et al., *Logic Synthesis for Finite State Machines Based on Linear Chains of States*, Studies in Systems, Decision and Control 113,
DOI 10.1007/978-3-319-59837-6_4

Table 4.1 Models of $P_C Y$ Moore FSMs

Model/Sources	PY	$P_C Y$	PY_1	$P_C Y_1$	PY_2	$P_C Y_2$	$P_C Y_3$
RG	•	○	○	•	•	○	•
BCT	○	•	○	•	○	•	•
BMO	○	○	•	○	•	•	•

$$B_i = \bigvee_{i=1}^{I} C_{mi} A_m. \tag{4.2}$$

In (4.2), C_{mi} is a Boolean variable equal to 1 if and only if (iff) $a_m \in B_i$. Let us encode the states $a_m \in A$ so that each function of system (4.2) is represented by minimum possible number of product terms. Such an encoding is named an optimal state assignment [3].

Let $\Pi_{RG} \subseteq \Pi_A$ be a set of classes $B_i \in \Pi_A$ such that each class is represented by a single interval of R-dimensional Boolean space. If $\Pi_{RG} = \Pi_A$, then the PY FSM should be used. Let us discuss the case when $\Pi_{RG} \neq \Pi_A$.

Let Π_{TC} be a set of classes $B_i \in \Pi_A$ required the transformation. Obviously, there is $\Pi_{TC} = \Pi_A \backslash \Pi_{RG}$. Let the set Π_{TC} include I_{TC} elements. To encode the classes $B_i \in \Pi_{TC}$, it is enough R_{TC} variables:

$$R_{TC} = \lceil \log_2 I_{TC} \rceil. \tag{4.3}$$

Let T_{EMB} be a set of possible amounts of EMB outputs. For up-to-day FPGAs, there is a set $T_{EMB} = \{1, 2, 4, 8, 16, 32, 64\}[1, 8]$. There are R inputs in the BMO. Therefore, a standard EMB should be configured in such a way that $V = 2^R$. After the configuration, this EMB block can implement microoperations $y_n \in Y$. It has t_F outputs, where t_F is the nearest number from T_{EMB} greater or equal to the value

$$t_0 = \left\lceil \frac{V_0}{2^R} \right\rceil. \tag{4.4}$$

The BMO generates N microoperations. The required number of EMBs in BMO is determined as

$$n_1 = \left\lceil \frac{N}{t_F} \right\rceil. \tag{4.5}$$

There are $n_1 \cdot t_F$ outputs in all EMBs forming the circuit of BMO. There are t_{BMO} "free" outputs which are not used for generating the microoperations. This value is determined as

$$t_{BMO} = n_1 t_F - N. \tag{4.6}$$

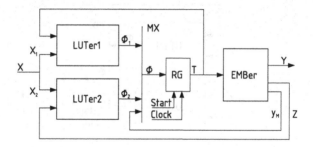

Fig. 4.1 Structural diagram of PY_2 Moore UFSM

Let the following condition take place:

$$t_{BMO} \geq R_{TC} + 1. \tag{4.7}$$

In this case, the block BTC is absent. Using the notation from Table 4.1, we can say that the condition (4.7) leads to PY_2 Moore UFSM (Fig. 4.1).

Now, the LUTer is represented by two blocks. The $LUTer_1$ implements the system

$$\Phi_1 = \Phi_1(T, X_1). \tag{4.8}$$

The block $LUTer_2$ implements the system

$$\Phi_2 = \Phi_2(Z, X_2). \tag{4.9}$$

The variables $z_r \in Z$ encode the classes $B_i \in \Pi_{TC}$. The set Z includes R_{TC} elements. It is quite possible that only some parts of the set X are used for generation functions (4.8) and (4.9). Obviously, there is $X_1 \cup X_2 = X$. In the general case, there is $X_1 \cap X_2 \neq \emptyset$. The choice of the state code is executed by the variable y_M using a multiplexer MX. For example, the code is determined by functions Φ_1 (Φ_1) if there is $y_M = 0$ ($y_M = 1$). Existance of y_M explains the second member in the right part of (4.7).

The EMBer generates microoperations Y and the following functions:

$$Z = Z(T); \tag{4.10}$$

$$y_M = y_M(T). \tag{4.11}$$

Let the following conditions take places:

$$\Pi_{RG} \neq \Pi_A; \tag{4.12}$$

$$t_{BMO} = 0. \tag{4.13}$$

It leads to $P_C Y_1$ Moore UFSM (Fig. 4.2). In this case, the functions (4.10)–(4.11) are generated by the block $LUTer_3$. A designer can use an additional EMB block to

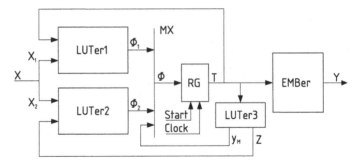

Fig. 4.2 Structural diagram of P_CY_1 Moore UFSM

Fig. 4.3 Structural diagram
of P_CY_2 Moore UFSM

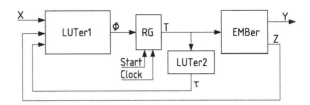

implement the functions (4.10)–(4.11). Of course, it leads to the transformation of P_CY_1 Moore UFSM into PY_2 Moore UFSM.

Let the following conditions take places:

$$\Pi_{RG} = \emptyset; \tag{4.14}$$

$$t_{BMO} \leq R_B + 1. \tag{4.15}$$

In this case, there are no class codes generated by RG. To eliminate MX, we propose to generate t_{BMO} bits of $K(B_i)$ by BMO. The rest of the code is generated by the block BCT (Fig. 4.3). We name the UFSM shown in Fig. 4.3 a P_CY_2 Moore UFSM.

In this model, the LUTer$_2$ implements functions

$$\tau = \tau(T). \tag{4.16}$$

These functions represent $(R_B - t_{BMO})$ bits of the code $K(B_i)$. The EMBer generates the microoperations Y represented by the system (4.16). Also it generates functions (4.10) representing t_{BMO} bits of the class codes.

Let the condition (4.15) take place, as well as the following conditions:

$$\Pi_{RG} = \emptyset; \tag{4.17}$$

$$\Pi_{TC} = \emptyset. \tag{4.18}$$

In this case, three sources of class codes should be used. It leads to P_CY_3 Moore UFSM (Fig. 4.4).

Fig. 4.4 Structural diagram
of $P_C Y_3$ Moore UFSM

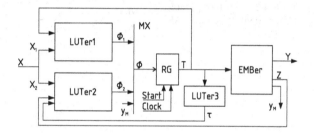

To simplify the block MX, functions τ and Z forming codes $K(B_i)$ of the classes $B_i \in \Pi_{TC}$ are used. The meaning is clear for all functions and variables shown in Fig. 4.4.

4.2   Design of UFSMs with Three Sources

Analysis of models proposed in Sect. 4.1 shows that the model $P_C Y_3$ possesses the most general nature. Let us discuss the design method for $P_C Y_3$ Moore UFSM. It includes the following steps:

1. nConstruction of partition $\Pi_A = \{B_1, \ldots, B_I\}$ of the set of states A by the classes of pseudoequivalent states.
2. Construction of the system $B(A)$.
3. Optimal state encoding targeted minimizing the number of terms in system $B(A)$. Construction of the set Π_{RG}.
4. Calculation of the values of t_{BMO} and R_{TC}. Construction of the sets τ and Z.
5. Encoding of the classes $B_i \in \Pi_{TC}$.
6. Construction of the table of LUTer1. Construction of the system Φ_1.
7. Construction of the table of LUTer2. Construction of the system Φ_2.
8. Construction of the table of LUTer3. Construction of the system τ.
9. Construction of the table of EMBer.
10. Implementing UFSM logic circuit with particular LUTs and EMBs.

Let us discuss an example of design for Moore UFSM $P_C Y_4(\Gamma_5)$. Let us start from the transformed table of transitions constructed on the base of some GSA Γ_5 (Table 4.2). This table determines transitions for some Moore FSM $PY(\Gamma_5)$. This table differs from a classical table of transitions because it represents the transitions for classes of PES [3]. As follows from Table 4.2, there are the following sets and parameters: $A = \{a_1, \ldots, a_{14}\}$, $M = 14$, $R = 4$, $T = \{T_1, \ldots, T_4\}$ and $\Phi = \{D_1, \ldots, D_4\}$.

Table 4.2 Transformed table of transitions of Moore FSM PY(Γ_5)

B_i	a_s	X_h	h
B_1	a_2	x_1	1
	a_3	$\bar{x}_1 x_2$	2
	a_5	$\bar{x}_1 \bar{x}_2$	3
B_2	a_1	x_3	4
	a_{10}	$\bar{x}_3$	5
B_3	a_4	x_2	6
	a_7	$\bar{x}_2 x_3$	7
	a_6	$\bar{x}_2 \bar{x}_3 x_4$	8
	a_{13}	$\bar{x}_2 \bar{x}_3 \bar{x}_4$	9
B_4	a_8	1	10
B_5	a_3	x_3	11
	a_9	$\bar{x}_3$	12
B_6	a_2	x_4	13
	a_{11}	$\bar{x}_4 x_5$	14
	a_{12}	$\bar{x}_4 \bar{x}_5 x_6$	15
	a_{14}	$\bar{x}_4 \bar{x}_5 \bar{x}_6$	16
B_7	a_1	$x_3 x_6$	17
	a_3	$x_3 \bar{x}_6$	18
	a_{10}	$x_2 \bar{x}_3$	19
	a_{12}	$\bar{x}_2 \bar{x}_3$	20

Let the following partition $\Pi_A = \{B_1, \ldots, B_7\}$ be constructed for the Moore FSM PY(Γ_5), where $B_1 = \{a_1\}$, $B_2 = \{a_5, a_{12}\}$, $B_3 = \{a_{11}, a_{13}, a_{14}\}$, $B_4 = \{a_3, a_6\}$, $B_5 = \{a_2, a_41\}$, $B_6 = \{a_7, a_8\}$, $B_7 = \{a_9, a_{10}\}$,. This partition can be represented by the following system $B(A)$:

$$
\begin{aligned}
B_1 &= A_1; \\
B_2 &= A_5; \\
B_3 &= A_{11} \vee A_{13} \vee A_{14}; \\
B_4 &= A_3 \vee A_6; \\
B_5 &= A_2 \vee A_4; \\
B_6 &= A_7 \vee A_8; \\
B_7 &= A_9 \vee A_{10}.
\end{aligned}
\tag{4.19}
$$

One of the variants of optimal state assignment is shown in Fig. 4.5.

Using the state codes from Fig. 4.5, the following system can be obtained for the initial system $B(A)$:

Fig. 4.5 Optimal state codes for Moore FSM $PY(\Gamma_5)$

T_1T_2 \ T_3T_4	00	01	11	10
00	a_1	a_3	a_6	*
01	a_{13}	a_{14}	*	a_5
11	a_{11}	a_9	a_8	a_{12}
10	a_7	a_2	a_4	a_{10}

$$B_1 = \bar{T}_1\bar{T}_2\bar{T}_4;$$
$$B_2 = T_2T_3\bar{T}_4;$$
$$B_3 = T_2\bar{T}_3\bar{T}_4 \vee \bar{T}_1T_2\bar{T}_3;$$
$$B_4 = \bar{T}_1\bar{T}_2T_4; \qquad (4.20)$$
$$B_5 = T_1\bar{T}_2T_4;$$
$$B_6 = T_1\bar{T}_2\bar{T}_3\bar{T}_4 \vee T_2T_3T_4;$$
$$B_7 = T_1T_2\bar{T}_3T_4 \vee \bar{T}_2T_3\bar{T}_4.$$

As follows from (4.20), the classes B_1, B_2, B_4, B_5 are represented by single generalized intervals of four-dimensional Boolean space. It gives the following sets of classes for Moore FSM $PY(\Gamma_5)$: $\Pi_{RG} = \{B_1, B_2, B_4, B_5\}$ and $\Pi_{TC} = \{B_3, B_6, B_7\}$.

Let it be the set $Y = \{y_1, \ldots, y_6\}$ for Moore FSM $PY(\Gamma_5)$. Let the following system of functions can be derived from GSA Γ_5:

$$y_1 = A_3 \vee A_5 \vee A_6 \vee A_{14};$$
$$y_2 = A_2 \vee A_7 \vee A_{11} \vee A_{13};$$
$$y_3 = A_2 \vee A_4 \vee A_8 \vee A_9 \vee A_{10};$$
$$y_4 = A_4 \vee A_5 \vee A_8 \vee A_{10} \vee A_{12}; \qquad (4.21)$$
$$y_5 = A_3 \vee A_6 \vee A_8 \vee A_9 \vee A_{14};$$
$$y_6 = A_2 \vee A_3 \vee A_4 \vee A_6 \vee A_7.$$

The system (4.21) is used for constructing a part of the table of EMBer.

Let us use an FPGA chip including EMBs with the following configurations: 128×1, 64×2, 32×4, 16×8, (bits). It gives the set $T_{EMB} = \{1, 2, 4, 8\}$, as well as the value $V_0 = 128$. In the case of $PY(\Gamma_5)$ there is $R = 4$. Using (4.4), the value $t_0 = 8$ can be found. Using (4.5), we can find the value $n_1 = 1$. Using (4.6), the value $t_{BMO} = 2$ can be found. It is necessary $R_{TC} = 2$ bits for encoding of the classes $B_i \in \Pi_{TC}$. So, the condition (4.7) is violated and the model P_CY_3 can be used. In the case of UFSM $P_CY_3(\Gamma_5)$, the following equality is true: $|\tau| = |z| = 1$.

Let us encode the classes $B_i \in \Pi_{TC}$ in the following way: $K(B_3) = 01$, $K(B_6) = 10$, and $K(B_7) = 00$. It gives the following equations:

Table 4.3 Table of LUTer$_1$ for Moore UFSM $P_C Y_3 (\Gamma_5)$

B_1	$K(B_i) T_1 T_2 T_3 T_4$	a_s	$K(a_s)$	X_h	Φ_h	h
B_1	$0 * 00$	a_2	1001	x_1	$D_1 D_4$	1
		a_3	0001	$\bar{x}_1 x_2$	D_4	2
		a_5	0110	$\bar{x}_1 \bar{x}_2$	$D_2 D_3$	3
B_2	$*110$	a_1	0000	x_3	$-$	4
		a_{10}	1010	$\bar{x}_3$	$D_1 D_3$	5
B_4	$00 * 1$	a_8	1111	1	$D_1 D_2 D_3 D_4$	6
B_5	$10 * 1$	a_3	0001	x_3	D_4	7
		a_9	1101	$\bar{x}_3$	$D_1 D_2 D_4$	8

$$\tau_1 = B_6 = T_1 \bar{T}_2 \bar{T}_3 \bar{T}_4 \vee T_2 T_3 T_4;$$
$$\tau_2 = B_3 = T_2 \bar{T}_3 \bar{T}_4 \vee \bar{T}_1 T_2 \bar{T}_3; \qquad (4.22)$$

To form the system (4.22), the equations from the system (4.20) are used.

The set τ includes variables implemented by the LUTer$_3$. So, those functions τ_r from (4.22) should be placed into τ whose circuits are implemented with the minimal amount of LUTs. Let us use LUTs having $S = 4$. It means that each of functions (4.22) is implemented using only single LUT. So, the sets τ and Z can be constructed in the arbitrary way. Let us form the following sets: $\tau = \{\tau_1\}$ and $Z = \{z_1\}$. So, the set τ includes the first bit of a class code, whereas the set Z contains the second.

The table of LUTer$_1$ includes transitions for classes $B_i \in \Pi_{RG}$ (Table 4.3).

This table is almost the same as a transformed structure table of Moore FSM. The class codes are taken from system (4.20). If some variable $T_r \in T$ is absent in the equation for $B_i \in \Pi_{RG}$, then it is represented by "$*$" in the code of $K(B_i)$. For example, there is no variable T_2 in the function B_1 (4.20). So, the code $0 * 00$ corresponds to the class $B_1 \in \Pi_{RG}$. As follows from Table 4.3, there is the set $X_1 = \{x_1, x_2, x_4\}$.

The LUTer$_1$ is characterized by the system (4.8). The following functions can be derived from Table 4.3:

$$D_1 = F_1 \vee F_5 \vee F_6 \vee F_8;$$
$$D_2 = F_3 \vee F_6 \vee F_8;$$
$$D_3 = F_3 \vee F_5 \vee F_6; \qquad (4.23)$$
$$D_4 = F_1 \vee F_2 \vee F_6 \vee F_7 \vee F_8.$$

The terms of this system are determined in the standard way. For example, $F_1 = \bar{T}_1 \bar{T}_3 \bar{T}_4 x_1$, $F_2 = \bar{T}_1 \bar{T}_3 \bar{T}_4 \bar{x}_1 x_2$, and so on.

The table of LUTer$_2$ includes transitions for classes $B_2 \in \Pi_{TC}$ (Table 4.4).

This table is constructed on the base of the table of transitions (or the structure table) of Moore FSM. The logic circuit of LUTer$_2$ is represented by the system (4.9). The following set $X_2 = \{x_2, \ldots, x_6\}$ can be derived from Table 4.4. The product terms

Table 4.4 Table of LUTer$_2$ for Moore UFSM $P_C Y_3(\Gamma_3)$

B_i	$K(B_i)\tau_1 z_1$	a_s	$K(a_s)$	X_h	Φ_h	h
B_3	00	a_4	1011	x_2	$D_1 D_3 D_4$	1
		a_7	1000	$\bar{x}_2 x_3$	D_1	2
		a_6	0011	$\bar{x}_2 \bar{x}_3 x_4$	$D_3 D_4$	3
		a_{13}	0100	$\bar{x}_2 \bar{x}_3 \bar{x}_4$	D_2	4
B_3	10	a_2	1001	x_4	$D_1 D_4$	5
		a_{11}	1000	$\bar{x}_4 x_5$	$D_1 D_2$	6
		a_{12}	1110	$\bar{x}_4 \bar{x}_5 x_6$	$D_1 D_2 D_3$	7
		a_{14}	0101	$\bar{x}_4 \bar{x}_5 \bar{x}_6$	$D_2 D_4$	8
B_7	11	a_1	0000	$x_3 x_6$	–	9
		a_3	0001	$x_3 \bar{x}_6$	D_4	10
		a_{10}	1010	$x_2 \bar{x}_3$	$D_1 D_3$	11
		a_{12}	1110	$\bar{x}_2 \bar{x}_3$	$D_1 D_2 D_3$	12

of the system (4.9) are determined by the rows of the table of LUTer$_2$. For example, the following terms can be found from Table 4.4: $F_1 = \tau_1 \bar{z}_1 x_2$, $F_2 = \tau_1 \bar{z}_2 \bar{x}_2 x_3$, and so on. The following system of input memory functions can be derived from Table 4.4:

$$
\begin{aligned}
D_1 &= F_1 \vee F_2 \vee F_5 \vee F_6 \vee F_7 \vee F_{11} \vee F_{12}; \\
D_2 &= F_4 \vee F_6 \vee F_7 \vee F_8 \vee F_{12}; \\
D_3 &= F_1 \vee F_3 \vee F_7 \vee F_{11} \vee F_{12}; \\
D_4 &= F_1 \vee F_3 \vee F_5 \vee F_8 \vee F_{10}.
\end{aligned}
\tag{4.24}
$$

The table of LUTer$_3$ is constructed to find the equations for functions $\tau_r \in \tau$. It includes the columns a_m, $K(a_m)$, B_i, $K(B_i)$, τ_h, h. It is constructed only for classes $B_i \in \Pi_{TC}$. In the discussed case, there is no need of this table. The equation for τ_1 has been already obtained. It is included into (4.22).

The table of EMBer includes the columns a_m, $K(a_m)$, $Y(a_m)$, y_m, Z. In the discussed case, it includes $M = 16$ rows (Table 4.5).

The variable $y_M = 0(1)$ for classes $B_i \in \Pi_{RG}$ ($B_i \in \Pi_{TC}$). The following equation determines the input memory functions Φ:

$$
\Phi = \bar{y}_M \Phi_1 \vee y_M \Phi_2.
\tag{4.25}
$$

This equation is used for implementing the circuit of the block MX.

The last step of design is reduced to implementing the logic circuit of UFSM. It is connected with application of some CAD tools such as WebPack [8]. We do not discuss this step in our book.

Table 4.5 Table of EMBer for Moore UFSM $P_C Y_3(\Gamma_5)$

a_m	$K(a_m)T_1T_2T_3T_4$	$Y(a_m)y_1y_2y_3y_4y_5y_6$	y_M	Zz_1
a_1	0000	000000	0	0
a_3	0001	100011	0	0
*	0010	000000	1	0
a_6	0011	100011	0	0
a_{13}	0011	010000	0	1
a_{14}	0101	100010	0	1
a_5	0110	100100	0	0
*	0111	000000	0	0
a_7	1000	010001	1	0
a_2	1001	011001	1	0
a_{10}	1010	001100	1	1
a_4	1011	001101	1	0
a_{11}	1100	010000	0	1
a_9	1101	001010	0	1
a_{12}	1110	000100	0	0
a_8	1111	001110	1	0

4.3   Design of UFSMs with Two Sources

Let us discus an example of design for UFSM $P_C Y_1(\Gamma_5)$. If an FPGA chip in use includes EMBs having the configuration 16×6, then the condition (4.13) takes place. Let us use the state codes from Fig. 4.5. It gives two sets of classes: $\Pi_{RG} = \{B_1, B_2, B_4, B_5\}$ and $\Pi_{TC} = \{B_3, B_6, B_7\}$. It is enough to have two variables for encoding the classes $B_i \in \Pi_{TC}$. It gives the set $Z = \{z_1, z_2\}$. Let an FPGA chip in use include LUTs having four inputs (S=4). Therefore, the following condition takes place:

$$S \geq R. \tag{4.26}$$

In this case, each function from system (4.10) is implemented using a single LUT. So, the class codes can be arbitrary. Let us use the following codes: $K(B_3) = 00$, $K(B_6) = 01$ and $K(B_7) = 10$.

Obviously, the tables of $LUTer_1$ are the same for UFSMs $P_C Y_3(\Gamma_5)$ and $P_C Y_1(\Gamma_5)$. The same is true for the systems Φ_1. The tables of $LUTer_2$ have equal structures but the codes are different in the column $K(B_i)$. In the case of UFSM $P_C Y_1(\Gamma_5)$, Table 4.6 represents the table of $LUTer_2$.

Obviously, the system Φ_2 is the same as (4.24). But it includes the following terms: $F_1 = \bar{z}_1\bar{z}_2x_2$, $F_2 = \bar{z}_1\bar{z}_2\bar{x}_2x_3$, and so on. The code 11 can be used for optimizing these terms. It leads, for example, to terms $F_5 = z_2x_4$, $F_{10} = z_1x_3\bar{x}_6$, and so on.

Table 4.6 Table of LUTer$_2$ for Moore UFSM $P_C Y_1(\Gamma_5)$

B_i	$K(B_i)$	a_s	$K(a_s)$	X_h	Φ_h	h
B_3	00	a_4	1011	x_2	$D_1 D_3 D_4$	1
		a_7	1000	$\bar{x}_2 x_3$	D_1	2
		a_6	0011	$\bar{x}_2 \bar{x}_3 x_4$	$D_3 D_4$	3
		a_{13}	0100	$\bar{x}_2 \bar{x}_3 \bar{x}_4$	D_2	4
B_6	01	a_2	1001	x_4	$D_1 D_4$	5
		a_{11}	1100	$\bar{x}_4 x_5$	$D_1 D_2$	6
		a_{12}	1100	$\bar{x}_4 \bar{x}_5 x_6$	$D_1 D_2 D_3$	7
		a_{14}	0101	$\bar{x}_4 \bar{x}_5 \bar{x}_6$	$D_1 D_4$	8
B_7	10	a_1	0000	$x_3 x_6$	–	9
		a_3	0001	$x_3 \bar{x}_6$	D_4	10
		a_{10}	1010	$x_2 \bar{x}_3$	$D_1 D_3$	11
		a_{12}	1110	$\bar{x}_2 \bar{x}_3$	$D_1 D_2 D_3$	12

Table 4.7 Table of LUTer$_3$ for Moore UFSM $P_C Y_1(\Gamma_5)$

a_m	$K(a_m)$	B_i	$K(B_i)$	Z_h	h
a_7	1000	B_6	01	z_2	1
a_8	1111	B_6	01	z_2	2
a_9	1101	B_7	10	z_1	3
a_{10}	1010	B_7	10	z_1	4

To find Boolean functions (4.10), the table of LUTer$_3$ should be constructed (Table 4.7).

There are no states $a_m \in B_3$ in Table 4.7. It has sense due to $K(B_3) = 00$. The following equations can be derived from Table 4.7:

$$z_1 = T_1 T_2 \bar{T}_3 T_4 \vee T_1 T_2 T_3 \bar{T}_4;$$
$$z_2 = T_1 \bar{T}_2 \bar{T}_3 \bar{T}_4 \vee T_1 T_2 T_3 T_4. \tag{4.27}$$

Let $y_M = 0$ for classes $B_i \in \Pi_{RG}$. It leads to the following equation:

$$y_M = \overline{A_1 \vee A_2 \vee A_3 \vee A_4 \vee A_5 \vee A_6 \vee A_{12}}. \tag{4.28}$$

Of course, this functions can be represented as:

$$y_M = A_7 \vee A_8 \vee A_9 \vee A_{10} \vee A_{11} \vee A_{13} \vee A_{14}. \tag{4.29}$$

Because the condition (4.26) takes place, each from Eqs. (4.28)–(4.29) is implemented using only a single LUT. If this condition is violated, the following equation can be used:

Fig. 4.6 Two-level structure of LUTer$_3$

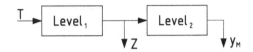

$$y_M = f(B_i) \ (B_i \in \Pi_{TC}). \tag{4.30}$$

In the discussed case, the following equation can be obtained:

$$y_M = B_3 \vee B_6 \vee B_7 = \bar{z}_1 \vee \bar{z}_2. \tag{4.31}$$

This approach results in two-level structure of LUTer$_3$ (Fig. 4.6).
Let the condition (4.12) take place, as well as the following condition:

$$t_{BMO} = 1. \tag{4.32}$$

In this case, the free output of EMBer can be used for generating the variable y_M. It changes only the table of EMBer for UFSM $P_C Y_1$.

In the discussed case, the condition (4.13) takes place. The tables of EMBer are practically the same for UFSMs $P_C Y_3(\Gamma_5)$ and $P_C Y_1(\Gamma_5)$. But columns y_M and z_1 are eliminated from Table 4.5 to construct the table of EMBer for $P_C Y_1(\Gamma_5)$.

Let the following condition be true:

$$t_{BMO} > 1. \tag{4.33}$$

In this case, the model $P_C Y_2$ can be used even if $\Pi_{RG} \neq \emptyset$. Let us discuss an example of design for UFSM $P_C Y_2(\Gamma_5)$. The proposed design method includes the following steps:

1. Construction of the partition Π_A.
2. State assignment.
3. Encoding of the classes $B_i \in Pi_A$.
4. Constructing the table of LUTer$_1$.
5. Constructing the table of LUTer$_2$.
6. Constructing the table of EMBer.
7. Implementing UFSM logic circuit.

The first step of the method is already executed. There is the partition $\Pi_A = \{B_1, \ldots, B_7\}$ with $I = 7$. Let us encode the states $a_m \in A$ in the trivial order: $K(a_1) = 0000, K(a_2) = 0001, \ldots, K(a_{14}) = 1101$. Using (3.16), the value of R_B can be found. Because of $I = 7$, there is $R_B = 3$. Let us encode the classes $B_i \in \Pi_A$ in the following manner: $K(B_1) = 000, K(B_2) = 001, \ldots, K(B_7) = 110$.

Let an FPGA in use have EMBs with the configuration 16×8. In the discussed case, it gives $t_{BMO} = 2$. Now, the following sets can be constructed: $\tau = \{\tau_1\}$ and $z = \{z_1, z_2\}$. The table of LUTer$_1$ can be constructed as an expansion of Table 4.2.

Table 4.8 Table of LUTer$_1$ for Moore UFSM $P_C Y_1(\Gamma_5)$

B_i	$K(B_i)\tau z_1 z_2$	a_s	$K(a_s)T_1 T_2 T_3 T_4$	X_h	Φ_h	h
B_1	000	a_2	0001	x_1	D_4	1
		a_3	0010	$\bar{x}_1 x_2$	D_3	2
		a_5	0100	$\bar{x}_1 \bar{x}_2$	D_2	3
B_2	001	a_1	0000	x_3	–	4
		a_{10}	1001	$\bar{x}_3$	$D_1 D_4$	5
B_3	010	a_4	0011	x_2	$D_3 D_4$	6
		a_7	0110	$\bar{x}_2 x_3$	$D_2 D_3$	7
		a_6	0101	$\bar{x}_2 \bar{x}_3 x_4$	$D_2 D_3$	8
		a_{13}	1100	$\bar{x}_2 \bar{x}_3 \bar{x}_4$	$D_1 D_2$	9
B_4	011	a_8	0111	1	$D_2 D_3 D_4$	10
B_5	100	a_3	0010	x_3	D_3	11
		a_9	1000	$\bar{x}_3$	D_1	12
B_6	101	a_2	0001	x_4	D_4	13
		a_{11}	1010	$\bar{x}_4 x_5$	$D_1 D_3$	14
		a_{12}	1011	$\bar{x}_4 \bar{x}_5 x_6$	$D_1 D_3 D_4$	15
		a_{14}	1101	$\bar{x}_4 \bar{x}_5 \bar{x}_6$	$D_1 D_2 D_4$	16
B_7	110	a_1	0000	$x_3 x_6$	–	17
		a_3	0010	$x_3 \bar{x}_6$	D_3	18
		a_{10}	1001	$x_2 \bar{x}_3$	$D_1 D_4$	19
		a_{12}	1011	$\bar{x}_2 \bar{x}_3$	$D_1 D_3 D_4$	20

It includes all columns of Table 4.2 and three new columns: $K(B_i)$, $K(a_s)$ and Φ_h (Table 4.8).

In $P_C Y_1$ UFSM, the LUTer1 implements the system of input memory functions

$$\Phi = \Phi(Z, \tau, X). \tag{4.34}$$

In the discussed case, the system (4.34) includes the following product terms: $F_1 = \bar{\tau}_1 \bar{z}_1 \bar{z}_2$, $F_1 = \bar{\tau}_1 \bar{z}_1 \bar{z}_2 \bar{x}_1 x_2, \ldots, F_1 = \tau_1 z_1 \bar{z}_2 \bar{x}_3 \bar{x}_2$. For example, the following Boolean function can be derived from Table 4.8:

$$D_2 = F_3 \vee [F_7 \vee F_8 \vee F_9] \vee [F_{10}] \vee [F_{16}]. \tag{4.35}$$

The expression $F_7 \vee F_8 \vee F_9$ can be optimized and represented as $\bar{\tau} z_1 \bar{z}_2 \bar{x}_2$. Taking into account the input assignment 111, the expressions for $[F_{10}]$ and $[F_{16}]$ can be simplified: $[F_{10}] = z_z z_2$; $[F_{16}] = \tau_1 z_2 \bar{x}_4 \bar{x}_5 \bar{x}_6$. Now, the Eq. (4.35) can be represented as the following:

$$D_2 = \bar{\tau}_1 \bar{z}_1 \bar{z}_2 \bar{x}_1 x_2 \vee \bar{\tau}_1 z_1 \bar{z}_2 \bar{x}_2 \vee z_1 z_2 \vee \tau_1 z_2 \bar{x}_4 x_5 \bar{x}_6. \tag{4.36}$$

Similar transformation can be done for all input memory functions $D_r \in \Phi$.

Table 4.9 Table of EMBer for Moore UFSM $P_C Y_1 (\Gamma_5)$

a_m	$K(a_m) T_1 T_2 T_3 T_4$	$Y(a_m) y_1 y_2 y_3 y_4 y_5 y_6$	$Z z_1 z_2$
a_1	0000	000000	00
a_2	0001	011001	00
a_3	0010	100011	11
a_4	0011	001101	00
a_5	0100	100100	01
a_6	0101	100011	11
a_7	0110	010001	01
a_8	0111	001110	01
a_9	1000	001010	10
a_{10}	1001	001100	10
a_{11}	1010	010000	10
a_{12}	1011	000100	01
a_{13}	1100	010000	10
a_{14}	1101	100010	10

To get an equation for $\tau_1 \in \tau$, it is necessary to find all class codes with the first position equal to 1. In the discussed case, the following classes can be found: B_5, B_6, B_7. It gives the following equation:

$$\tau_1 = A_2 \vee A_4 \vee A_7 \vee A_8 \vee A_9 \vee A_{10}. \qquad (4.37)$$

This equation is used for implementing the circuit of LUTer2. Let $S = 4$, then this block is implemented by a single LUT.

The table of EMBer is constructed in the trivial way. It includes the columns: a_m, $K(a_m)$, $Y(a_m)$, Z. In the discussed case, it includes 16 rows (Table 4.9).

Let us explain the column Z. For example, the class $B_2 \in \Pi_A$ includes the states $a_5, a_{12} \in A$. Because of $K(B_2) = 001$, the rows a_5 and a_{12} contain 01 in the column Z. This principle is used for filling all rows a_m in Table 4.9.

Let us point out that the model $P_C Y_1$ does not include the block MX. So, it can be expected that it has the minimum propagation time among all discussed models. But in general case, equations are more complex for functions $D_r \in \Phi$ of $P_C Y_1$. It means that the final conclusion can be done after the implementation of the UFSM logic circuit. But the $P_C Y_1$ UFSM can be considered as an alternative for other structures if the condition (4.13) has no place.

Let us point out that the state assignment can be executed so as to optimize the circuit of LUTer1. Let us discuss the following variant of state assignment (Fig. 4.7).

As follows from the Karnaugh map (Fig. 4.7), the Eq. (4.37) can be represented as:

$$\tau_1 = T_1. \qquad (4.38)$$

It means that circuit of LUTer2 is reduced up to a wire (Fig. 4.8).

Fig. 4.7 Variant of state
assignment for Moore
UFSM $P_C Y_1(\Gamma_5)$

T_3T_4 \ T_1T_2	00	01	11	10
00	a_1	a_3	a_2	a_4
01	a_5	a_6	a_7	a_8
11	a_{13}	a_{14}	a_9	a_{10}
10	a_{11}	a_{12}	*	*

Fig. 4.8 Structural diagram
of $P_C Y_1(\Gamma_5)$ with
optimizing the LUTer2

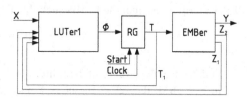

4.3.1   Synthesis of UFSMs with Transformation of Microoperations

This method is based on a transformation of microoperations $y_n \in Y$ into the codes of classes of PES $B_i \in \Pi_A$. It is one of the methods that belongs to the class of object transformation methods [2]. We discuss this approach in details in Chap. 5.

Let Q be a number of different collections of microoperations $Y_q \subseteq Y$ for a GSA Γ. Obviously, the following condition takes place:

$$Q \le M. \tag{4.39}$$

Let M_q be a number of states $a_m \in A$ including a CMO Y_q ($q = 1, \ldots, Q$). Let us find the value

$$M_{\max} = \max(M_1, \ldots, M_Q). \tag{4.40}$$

This parameter is equal to the cardinality number of the set IS including identifiers of states $a_m \in A$. Now each state $a_m \in A$ can be represented as the following vector

$$a_m = \langle I_m, Y(a_m) \rangle \ (m = 1, \ldots, M). \tag{4.41}$$

If $Y(a_m)$ is a CMO Y_q such that $M_q = 1$, then $I_m = I_1$. The identifiers should be different for states $a_m, a_s \in A$ if $Y(a_m) = Y(a_s)$. Let us encode identifiers $I_m \in IS$ by binary codes $K(I_m)$ having R_I bits:

$$R_I = \lceil \log_2 M_{\max} \rceil. \tag{4.42}$$

Fig. 4.9 Structural diagram
of $P_Y Y$ Moore UFSM

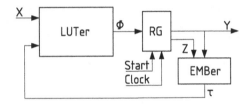

Let us use the variables $z_r \in Z$ for encoding of identifiers, where $|Z| = R_I$.

Let us encode the classes $B_i \in \Pi_A$ by binary codes $K(B_i)$ having R_B bits. Let an FPGA chip in use include EMBs such that the following condition is true:

$$V_0 \geq 2^{N+R_I} \cdot R_B. \tag{4.43}$$

In this case, the structural diagram of $P_Y Y$ Moore UFSM is proposed (Fig. 4.9).

In $P_Y Y_F SM$, the set Φ includes $N + R_I$ elements. The LUTer implements the system of input memory functions $\Phi = \Phi(\tau, x)$ Two sets of variables are generated by RG, namely, the microoperations $y_n \in Y$ and the additional variables $z_r \in Z$. The EMBer implements the system.

$$\tau = \tau(Z, Y). \tag{4.44}$$

The proposed design method for $P_Y Y$ Moore UFSM includes the following steps:

1. Marking the initial GSA Γ by states of Moore FSM.
2. Constructing the collections of microoperations.
3. Constructing the set of identifiers.
4. Representing the states by pairs $\langle$ identifier, CMO $\rangle$.
5. Constructing the partition Π_A.
6. Encoding of the classes of PES.
7. Encoding of the identifiers $I_m \in IS$.
8. Constructing the table of LUTer.
9. Constructing the table of EMBer.
10. Implementing the UFSM logic circuit.

Let us discuss an example of design for $P_Y Y(\Gamma_6)$ Moore UFSM. The marked GSA Γ_6 is shown in Fig. 4.10.

The following sets and their parameters can be found from GSA Γ_6: $X = \{x_1, \ldots, x_5\}$, $Y = \{y_1, \ldots, x_6\}$, $A = \{a_1, \ldots, a_{10}\}$, $L = 5$, $N = 6$, $M = 10$. Therefore, there is $R = 4$ and it defines the sets $T = \{T_1, \ldots, T_4\}$ and $\Phi = \{D_1, \ldots, D_4\}$.

There are $Q = 7$ different collections of microoperations in the discussed case: $Y_1 = \emptyset$, $Y_2 = \{y_1, y_2\}$, $Y_3 = \{y_2, y_4\}$, $Y_4 = \{y_3\}$, $Y_5 = \{y_1, y_4, y_5\}$, $Y_6 = \{y_5, y_6\}$, $Y_7 = \{y_3, y_5\}$. Let us find the values of M_q. Analysis of GSA Γ_6 gives the following values: $M_q = 2$ for $q \in \{2, 3, 4\}$ and $M_q = 1$ for other CMOs. It means that $M_{max} = 2$ and $IS = \{I_1, I_2\}$. It gives $R_I = 1$ and $Z = \{z_1\}$. Now, the following pairs (4.41) can be constructed for states $a_m \in A$ (Table 4.10).

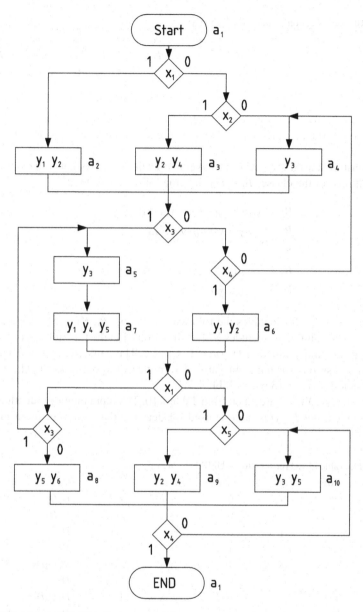

Fig. 4.10 Initial marked GSA Γ_6

The following partition Π_A can be found for the set A: $\Pi_A = \{B_1, \ldots, B_5\}$. It includes the classes $B_1 = \{a_1\}$, $B_2 = \{a_2, a_3, a_4\}$, $B_3 = \{a_5\}$, $B_4 = \{a_6, a_7\}$, $B_5 = \{a_8, a_9, a_{10}\}$. So, there is $I = 5$. It determines the value $R_B = 3$ and the set $\tau = \{\tau_1, \tau_2, \tau_3\}$. Let us encode the classes $B_i = \Pi_A$ in the following manner: $K(B_1) =$

Table 4.10 Identification of states for $P_Y Y(\Gamma_6)$

a_m	a_1	a_2	a_3	a_4	a_5	a_6	a_7	a_8	a_9	a_{10}
I_m	I_1	I_1	I_1	I_1	I_2	I_2	I_1	I_1	I_2	I_1
$Y(a_m)$	Y_1	Y_2	Y_3	Y_4	Y_4	Y_2	Y_5	Y_6	Y_3	Y_7

000, $K(B_2) = 001, \ldots, K(B_5) = 100$. Let us encode the identifiers: $K(I_1) = 0$ and $K(I_2) = 1$.

To construct the table of LUTer, let us construct the system of generalized formulae of transitions for the classes $B_i \in \Pi_A$. It is the following system:

$$
\begin{aligned}
B_1 &= x_1 a_2 \vee \bar{x}_1 x_2 a_3 \vee \bar{x}_1 \bar{x}_2 a_4; \\
B_2 &= x_3 a_5 \vee \bar{x}_3 x_4 a_6 \vee \bar{x}_3 \bar{x}_4 a_4; \\
B_3 &= a_7; \\
B_4 &= x_1 x_3 a_5 \vee x_1 \bar{x}_3 a_8 \vee \bar{x}_1 x_5 a_9 \vee \bar{x}_1 \bar{x}_5 a_{10}; \\
B_5 &= x_4 a_1 \vee \bar{x}_4 a_{10}.
\end{aligned}
\tag{4.45}
$$

The table of LUTer includes the following columns: B_i, $K(B_i)$, I_S, $K(I_S)$, $Y(a_s)$, X_h, Φ_h, h. The column Φ_h includes $N + R_I$ functions. Let the function y_n corresponds to the input memory function D_n $(n = 1, \ldots, N)$. The pairs $\langle I_S, Y(a_5)\rangle$ determines states of transitions for the right parts of formulae of system (4.45). The table of LUTer includes $H_0 = 13$ rows (Table 4.11).

The pairs $\langle I_S, Y(a_5)\rangle$ are taken from Table 4.10. The input memory function $D_7 = 1$ for the identifier I_2. This table is used for deriving the system of input memory

Table 4.11 Table of LUTer for Moore UFSM $P_Y Y(\Gamma_6)$

B_i	$K(b_i)$	I_s	$K(I_s)$	$Y(a_s)$	X_h	Φ_h	h
B_1	000	I_1	0	$y_1 y_2$	x_1	$D_1 D_2$	1
		I_1	0	$y_2 y_4$	$\bar{x}_1 x_2$	$D_2 D_4$	2
		I_1	0	y_3	$\bar{x}_1 \bar{x}_2$	D_3	3
B_2	001	I_2	1	y_3	x_3	$D_3 D_7$	4
		I_2	1	$y_1 y_2$	$\bar{x}_3 x_4$	$D_1 D_2 D_7$	5
		I_1	0	y_3	$\bar{x}_3 \bar{x}_4$	D_3	6
B_3	010	I_1	0	$y_1 y_4 y_5$	1	$D_1 D_4 D_5$	7
B_4	011	I_2	1	y_3	$x_1 x_3$	$D_3 D_7$	8
		I_1	0	$y_5 y_6$	$x_1 \bar{x}_3$	$D_5 D_6$	9
		I_2	1	$y_2 y_4$	$\bar{x}_1 x_5$	$D_2 D_4 D_7$	10
		I_1	0	$y_3 y_5$	$\bar{x}_1 \bar{x}_5$	$D_3 D_5$	11
B_5	100	I_1	0	–	x_4	–	12
		I_1	0	$y_3 y_5$	$\bar{x}_4$	$D_3 D_5$	13

functions. For example, the following expression can be found for function D_1:

$$D_1 = F_1 \lor F_5 \lor F_7. \tag{4.46}$$

The terms F_h in (4.46) are determined as the following conjunctions: $F_1 = \bar{\tau}_1 \bar{\tau}_2 \bar{\tau}_3 x_1$; $F_5 = \bar{\tau}_1 \bar{\tau}_2 \bar{\tau}_3 \bar{x}_3 x_4$; $F_7 = \bar{\tau}_1 \tau_2 \bar{\tau}_3$. There are unused input assignments of variables $\tau_r \in \tau$, such as 101, 110 and 111. They can be used for simplifying terms F_5 (the input assignment 101) and F_7 (the input assignment 110). It gives the following conjunctions $F_5 = \bar{\tau}_2 \tau_3 \bar{x}_3 x_4$ and $F_7 = \tau_2 \bar{\tau}_3$.

The table of EMBer is constructed on the base of the following matching:

$$\langle I_M, Y(G_m) \rangle \rightarrow (B_i | a_m \in B_i). \tag{4.47}$$

The table of EMBer includes the following columns: a_m, B_i, I_m, $Y(a_m)$, $K(I_m)$, $K(B_i)$, m. The columns $Y(a_m)$, $K(I_m)$ form the address of cell number m. The column $K(B_i)$ gives the content of the cell. In the general case, this table includes $H(\mathrm{P_Y}Y)$ rows, where

$$H(\mathrm{P_Y}Y) = 2^{N+R_I}. \tag{4.48}$$

In the discussed case, the expression (4.48) gives 128 rows. Only 10 of them are used for implementing the function (4.44). These rows are represented by Table 4.12.

The main advantage of the model $\mathrm{P_Y}Y$ is the highest possible performance. There is only a single level of logic for generating microoperations. But this model can be used only if the condition (4.43) takes place. If this condition is violated, we propose to encode the collections of microoperations [3]. It leads to $\mathrm{P_Y}Z$ model of Moore FSM (Fig. 4.11).

In $\mathrm{P_Y}Z$FSM, the set Φ includes $R_I + R_Y$ elements, where the value of R_Y is determined as

$$R_Y = \lceil \log_2 Q \rceil. \tag{4.49}$$

Table 4.12   Part of table of EMBer for Moore UFSM $\mathrm{P_Y}Y(\Gamma_6)$

a_m	B_i	I_m	$Y(a_m)y_1y_2y_3y_4y_5y_6$	$K(I_m)z_1$	$K(B_i)\tau_1\tau_2\tau_3$	m
a_1	B_1	I_1	000000	0	000	1
a_2	B_2	I_1	110000	0	001	2
a_3	B_2	I_1	010100	0	001	3
a_4	B_2	I_1	001000	0	001	4
a_5	B_3	I_2	001000	1	010	5
a_6	B_4	I_2	110000	1	011	6
a_7	B_4	I_1	100110	0	011	7
a_8	B_5	I_1	000011	0	100	8
a_9	B_5	I_2	010100	1	100	9
a_{10}	B_5	I_1	001010	0	100	10

Fig. 4.11 Structural
diagram of $P_Y Z$ Moore FSM

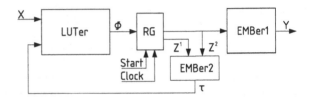

The variables $z_r \in Z^1$ are used for encoding of identifiers $I_m \in IS$, where $|Z^1| = R_I$. The variables $z_r \in Z^2$ are used for encoding of the collections of microoperations $Y_q \subseteq Y$, $D_r \in \Phi$, where $|\Phi| = R_I + R_Y$. The EMBer1 implements the microoperations $y_n \in Y$ represented as

$$Y = Y(Z^2). \tag{4.50}$$

The EMBer2 implements the variables $\tau_r \in \tau$ used for encoding of the classes $B_i \in \Pi_A$. So, it implements the system

$$\tau = \tau(Z^1, Z^2). \tag{4.51}$$

The design method for $P_Y Z$ UFSM can be obtained from the previous one. To do it, the step 9 is replaced by the steps 9^a and 9^b:

9^a. Constructing the table of EMBer1.
9^b. Constructing the table of EMBer2.

Also, the step 7^a is introduced and executed after the step 7:

7^a. Encoding of the collections of microoperations.

Let us discuss an example of design for the Moore UFSM $P_Y Z(\Gamma_6)$. The steps from 1 to 7 have been already executed. Let us start from the step 7^a.

Because of $Q = 7$, there is $R_Y = 3$. It gives the set $Z^2 = \{z_1, z_2, z_3\}$. Let us point out that $Z^1 = \{z_4\}$. Let us encode the CMOs in the following manner: $K(Y_1) = 000$, $K(Y_2) = 001$, and so on.

The table of LUTer includes the following columns: B_i, $K(B_i)$, I_S, $K(I_S)$, Y_h, $K(Y_h)$, X_h, Φ_h, h. The column Φ_h includes $R_I + R_Y = 4$ variables $D_r \in \Phi$. Let the variables $D_1 - D_3$ load the code $K(Y_q)$ into RG, whereas the variable D_4 the code $K(I_S)$. The table is constructed using Table 4.10 and system (4.45). As in previous case, it includes 13 rows (Table 4.13).

This table is used for deriving the system of input memory functions. For example the following equation can be derived: $D_3 = F_1 \vee F_3 \vee [F_4 \vee F_6] \vee [F_8 \vee F_9]$. After using the law of expansion [4], the following equation can be obtained:

$$D_3 = \bar{\tau}_1 \bar{\tau}_2 \bar{\tau}_3 x_1 \vee \bar{\tau}_1 \bar{\tau}_2 \bar{\tau}_3 \bar{x}_3 \vee \bar{\tau}_1 \tau_2 \tau_3 x_1. \tag{4.52}$$

Table 4.13 Table of LUTer for Moore UFSM $P_Y Z(\Gamma_6)$

B_i	$K(B_i)$	I_s	$K(I_s)$	Y_h	$K(Y_h)$	X_h	Φ_h	h
B_1	000	I_1	0	Y_2	001	x_1	D_3	1
		I_1	0	Y_3	010	$\bar{x}_1 x_2$	D_2	2
		I_1	0	Y_4	011	$\bar{x}_1 \bar{x}_2$	$D_2 D_3$	3
B_2	001	I_2	1	Y_4	100	x_3	$D_1 D_4$	4
		I_2	1	Y_2	001	$\bar{x}_3 x_4$	$D_3 D_4$	5
		I_1	0	Y_4	011	$\bar{x}_3 \bar{x}_4$	$D_2 D_3$	6
B_3	010	I_1	0	Y_5	100	1	D_1	7
B_4	011	I_2	1	Y_4	011	$x_1 x_3$	$D_1 D_3 D_4$	8
		I_1	0	Y_6	101	$x_1 \bar{x}_3$	$D_1 D_3$	9
		I_2	1	Y_3	010	$\bar{x}_1 x_5$	$D_2 D_4$	10
		I_1	0	Y_7	110	$\bar{x}_1 \bar{x}_5$	$D_1 D_2$	11
B_5	100	I_1	0	Y_1	000	x_4	–	12
		I_1	0	Y_7	110	$\bar{x}_4$	$D_1 D_2$	13

Table 4.14 Table of EMBer1 for Moore UFSM $P_Y Z(\Gamma_6)$

$K(Y_q) z_1 z_2 z_3$	$Y_q y_1 y_2 y_3 y_4 y_5 y_6$	q
000	000000	1
001	110000	2
010	010100	3
011	001000	4
100	100110	5
101	000011	6
110	001010	7
111	000000	*

Similar Boolean functions can be derived for all functions $D_r \in \Phi$. The table of EMBer1 is constructed in the trivial way. It includes the columns $K(Y_q)$, Y_q, q (Table 4.14).

The table of EMBer2 is constructed on the base of the matching (4.47). It includes the columns: a_m, B_i, I_m, $Y(a_m)$, $K(Y(a_m))$, $K(I_m)$, $K(B_i)$, m. The columns $K(Y(a_m))$, $K(I_m)$ form addresses of cells. This table includes $H(P_Y Z)$ rows, where

$$H(P_Y Z) = 2^{R_Y + R_I}. \tag{4.53}$$

In the discussed case, the expression (4.53) gives 16 rows. Only 10 of them are used for implementing the functions of the system (4.51). If we compare blocks EMBer of $P_Y Y(\Gamma_6)$ and EMBer2 of $P_Y Z(\Gamma_6)$, we can see that the number of required cells is diminished in 8 times. The table of EMBer2 is shown in Table 4.15. Only 10 rows are shown in this table.

Table 4.15 Table of EMBer2 for Moore UFSM $P_Y Z(\Gamma_6)$

a_m	B_i	I_m	$Y(a_m)$	$K(Y(a_m))$	$K(I_m)$	$K(B_i)$	m
a_1	B_1	I_1	Y_1	000	0	000	1
a_2	B_2	I_1	Y_2	001	0	001	2
a_3	B_2	I_1	Y_3	010	0	001	3
a_4	B_2	I_1	Y_4	011	0	001	4
a_5	B_3	I_2	Y_4	011	1	010	5
a_6	B_4	I_2	Y_1	000	1	011	6
a_7	B_4	I_1	Y_5	100	0	011	7
a_8	B_5	I_1	Y_6	101	0	100	8
a_9	B_5	I_2	Y_3	010	1	100	9
a_{10}	B_5	I_1	Y_7	110	0	100	10

Fig. 4.12 Structural diagram of $P_Y Z_0$ Moore UFSM

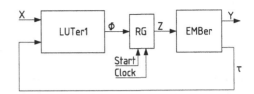

Let the following condition take place:

$$V_0 \geq 2^{R_Y + R_I}(N + R_B). \tag{4.54}$$

In this case, both blocks EMBer1 and EMBer2 can be combined in a single block EMBer. It leads to the model $P_Y Z_0$ shown in Fig. 4.12.

In $P_Y Z_0$ Moore UFSM, the set $Z = Z^1 \cup Z^2$. The table of EMBer can be used for implementing FSM logic circuits. Three different elements can be used for implementing an FSM logic circuit, namely, LUTs, EMBs and PLAs. The PLA blocks can be used, for example, for implementing logic circuit of EMBer in $P_Y Y$ Moore FSM. We do not discuss that issue in this Chapter.

4.4 Replacement of Logical Conditions

The replacement of logical conditions [3] is an universal method targeting the hardware reduction of BIMF. It can be used in all UFSMs discussed in this Chapter. An additional block of replacement of logical conditions (BRLC) should be introduced. It implements the system (2.19). Of course, the terms of this system can be different for different models of UFSMs. Existence of BRLC is devoted by the letter "M" in the name of a model.

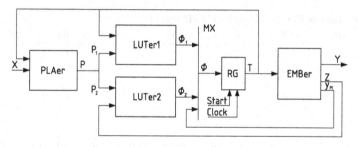

Fig. 4.13 Structural diagram of MPY$_2$ Moore UFSM

Let us start from MPY$_2$ Moore UFSM. Let PLAs be used for implementing the BRLC logic circuit. Let the following condition take place:

$$S \geq L + R. \tag{4.55}$$

In this case, the following model of MPY$_2$ Moore UFSM is proposed (Fig. 4.13).

A block PLAer consists from PLA-cells. If the condition (4.55) is true, then only one PLA is necessary to implement the PLAer. It implements the system (2.19). The block LUTer1 implements the system

$$\Phi_1 = \Phi_1(T, P_1). \tag{4.56}$$

The block LUTer2 implements the system

$$\Phi_2 = \Phi_2(\tau, P_2). \tag{4.57}$$

The functions of EMBer are the same as for PY$_2$ Moore UFSM.

The proposed design method for MPY$_2$ Moore UFSM includes the following steps:

1. Marking the initial GSA by states of Moore FSM.
2. Constructing the partition Π_A for the set A.
3. Constructing the system B(A).
4. Executing the optimal state assignment.
5. Finding the sets Π_{RG}, Π_{TC} and Z.
6. Replacing the logical conditions.
7. Constructing the table of PLAer.
8. Constructing the table of LUTer1.
9. Encoding the classes $B_i \in \Pi_{TC}$.
10. Constructing the table of LUTer2.
11. Constructing the table of EMBer.
12. Implementing the logic circuit of UFSM.

Table 4.16 Table of RLC for Moore UFSM $MPY_2(\Gamma_5)$

B_i	B_1	B_2	B_3	B_4	B_5	B_6	B_7
p_1	x_2	–	x_2	–	–	x_5	x_2
p_2	x_1	x_3	x_3	–	x_3	x_6	x_3
p_3	–	–	x_4	–	–	x_4	x_6

Let us discuss an example of synthesis for the Moore UFSM $MPY_2(\Gamma_5)$. The steps from 1 to 5 are already executed. The following sets are found: $A = \{a_1, \ldots, a_{14}\}$, $\Pi_A = \{B_1, \ldots, B_7\}$, $B_1 = \{a_1\}$, $B_2 = \{a_5, a_{12}\}$, $B_3 = \{a_{11}, a_{13}, a_{14}\}$, $B_4 = \{a_3, a_6\}$, $B_5 = \{a_2, a_4\}$, $B_6 = \{a_7, a_8\}$, $B_7 = \{a_9, a_{10}\}$, $\Pi_{RG} = \{B_1, B_2, B_4, B_5\}$, $\Pi_{TC} = \{B_3, B_6, B_7\}$. The optimal state codes can be taken from Fig. 4.5.

Let us denote the set of logical conditions determining transitions from states $a_m \in B_i$ as $X(B_i)$ $(i = 1, \ldots, I)$. The $X(B_2) = \{x_3\}$, $X(B_3) = \{x_2, x_3, x_4\}$, $X(B_4) = \emptyset$, $X(B_5) = \{x_3\}$, $X(B_6) = \{x_4, x_5, x_6\}$, $X(B_7) = \{x_2, x_3, x_6\}$. Therefore, there is $G = 3$ and $P = \{p_1, p_2, p_3\}$. Let $G_1(G_2)$ be the number of additional variables determining transitions from the states for LUTer1(LUTer2). In the discussed case, there are $G_1 = 2$ and $G_2 = 3$. It gives the sets $X_1 = \{x_1, x_2, x_3\}$ and $X_2 = \{x_2, x_3, x_4, x_5, x_6\}$.

Let us place the conditions $x_l \in X_1$ only in the column p_1, p_2 of the table of replacement of logical conditions. It leads to Table 4.16.

The following equations can be derived from Table 4.16:

$$
\begin{aligned}
p_1 &= (B_1 \vee B_3 \vee B_7) \vee B_6 x_5; \\
p_2 &= B_1 x_1 \vee (B_2 \vee B_3 \vee B_5 \vee B_7) x_3 \vee B_6 x_6; \\
p_3 &= (B_3 \vee B_6) x_4 \vee B_7 x_6;
\end{aligned}
\tag{4.58}
$$

The system (4.57) is implemented using the PLAer. The terms from the system (4.20) are used for representing the classes of PES in (4.57). So, Table 4.16 replaces the table of PLAer.

The table of LUTer1 has the same columns as its counterpart from the UFSM $PY_2(\Gamma_5)$. But the column X_h is replaced by the column P_h (Table 4.17).

Let us use the same class codes for $B_i \in \Pi_{TC}$ as it is for the UFSM $PY_2(\Gamma_5)$: $K(B_3) = 00$, $K(B_6) = 01$ and $K(B_7) = 10$. Now, the table of LUTer2 can be constructed. It includes the same columns as its counterpart for the UFSM $PY_2(\Gamma_5)$. But the column X_h is replaced by the column P_h (Table 4.18).

The content of EMBer does not depend on outcome of RLC. So, this table is the same as its counterpart for UFSM $PY_2(\Gamma_5)$. We leave this task to a reader.

The functions (4.56) can be derived from Table 4.17. They have the same structure as the functions from (4.23). But they depend on different product terms. For example, there are the terms $F_1 = \bar{T}_1 \bar{T}_3 \bar{T}_4 p_2$, $F_1 = \bar{T}_1 \bar{T}_3 \bar{T}_4 \bar{p}_2$ and so on. The same is true for functions (4.57). They are derived from Table 4.18 and have the same structure as the functions of (4.24). But they include different product terms. For example, there are the terms $F_1 = \bar{z}_1 \bar{z}_2 p_1$, $F_2 = \bar{z}_1 z_2 \bar{p}_1 p_2$, $F_3 = \bar{z}_1 z_2 \bar{p}_1 \bar{p}_3$, and so on.

Table 4.17 Table of LUTer1 for Moore UFSM $MPY_2(\Gamma_5)$

B_i	$K(B_i)$	a_s	$K(a_s)$	P_h	Φ_h	h
B_1	0*00	a_2	1001	p_2	D_1D_4	1
		a_3	0001	$p_1\bar{p}_2$	D_4	2
		a_5	0110	$\bar{p}_1\bar{p}_2$	D_2D_3	3
B_2	*110	a_1	0000	p_2	–	4
		a_{10}	1010	$\bar{p}_2$	D_1D_3	5
B_4	00*1	a_8	1111	1	$D_1D_2D_3D_4$	6
B_5	10*1	a_3	0001	p_2	D_4	7
		a_9	1101	$\bar{p}_2$	$D_1D_2D_4$	8

Table 4.18 Table of LUTer2 for Moore UFSM $MPY_2(\Gamma_5)$

B_i	$K(B_i)$	a_s	$K(a_s)$	P_h	Φ_h	h
B_3	00	a_4	1011	p_1	$D_1D_3D_4$	1
		a_7	1000	$\bar{p}_1p_2$	D_1	2
		a_6	1011	$\bar{p}_1\bar{p}_2p_3$	D_3D_4	3
		a_{13}	0100	$\bar{p}_1\bar{p}_2\bar{p}_3$	D_2	4
B_6	01	a_2	1001	p_3	D_1D_4	5
		a_{11}	1100	$p_1\bar{p}_3$	D_1D_2	6
		a_{12}	1110	$\bar{p}_1p_2\bar{p}_3$	$D_1D_2D_3$	7
		a_{14}	0101	$\bar{p}_1\bar{p}_2\bar{p}_3$	D_2D_4	8
B_7	10	a_1	0000	p_2p_3	–	9
		a_3	0001	$p_2\bar{p}_3$	D_4	10
		a_{10}	1010	$p_1\bar{p}_2$	D_1D_3	11
		a_{12}	1110	$\bar{p}_1\bar{p}_2$	$D_1D_2D_2D_3$	12

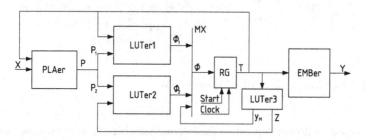

Fig. 4.14 Structural diagram of MP_CY_1 Moore UFSM

Using the same approach, the following models can be proposed: MP_CY_1 (Fig. 4.14), MP_CY_2 (Fig. 4.15), MP_CY_3 (Fig. 4.16), MP_YY (Fig. 4.17) and MP_YZ (Fig. 4.18).

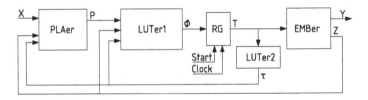

Fig. 4.15 Structural diagram of $MP_C Y_2$ Moore UFSM

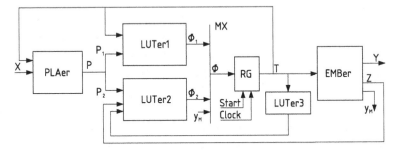

Fig. 4.16 Structural diagram of $MP_C Y_3$ Moore UFSM

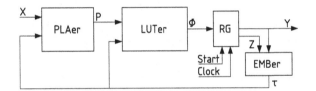

Fig. 4.17 Structural diagram of $MP_Y Y$ Moore UFSM

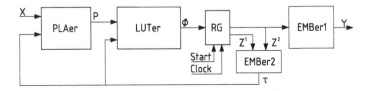

Fig. 4.18 Structural diagram of $MP_Y Z$ Moore UFSM

If functions P do not depend on the internal variables $T_r \in T$, then the circuit of PLAer can be optimized. To do it, we propose the method of optimal class assignment. Let us encode the classes $B_i \in \Pi_A$ as it is shown in Fig. 4.19. For example, they can be used in $MP_Y Y (\Gamma_5)$.

Using these codes and the system (4.58), the following system can be obtained:

$$
\begin{aligned}
p_1 &= \bar{\tau}_1 x_2 \vee \tau_2 \bar{\tau}_3 x_5; \\
p_2 &= \bar{\tau}_2 \bar{\tau}_3 x_1 \vee \tau_3 x_3 \vee \tau_2 \bar{\tau}_3 x_6; \\
p_3 &= (\bar{\tau}_1 \tau_2 \vee \tau_2 \bar{\tau}_3) x_4 \vee \bar{\tau}_1 \bar{\tau}_2 \tau_3 x_6.
\end{aligned}
\tag{4.59}
$$

Fig. 4.19 Optimal class
codes for UFSM $MP_Y Y (\Gamma_5)$

$\tau_1 \backslash \tau_2 \tau_3$	00	01	11	10
0	B_1	B_7	B_3	*
1	*	B_5	B_2	B_6

The system (4.58) includes 13 terms, whereas the system (4.59) only 8. It reduces the requirements to the number of terms in PLA cells.

References

1. Altera Corporation Home Page. http://www.altera.com
2. Barkalov, A., Barkalov Jr., A.: Design of mealy finite state machines with the transformation of object codes. Int. J. Appl. Math. Comput. Sci. **15**(1), 151–158 (2005)
3. Barkalov, A., Titarenko, L.: Logic Synthesis for FSM-Based Control Units. Lecture Notes in Electrical Engineering. Springer, Berlin (2009)
4. Barkalov, A., Titarenko, L.: Basic Principles of Logic Design. UZ Press, Zielona Góra (2010)
5. Barkalov, A., Titarenko, L., Chmielewski, S.: Reduction in the number of pal macrocells in the circuit of moore fsm. Int. J. Appl. Math. Comput. Sci. **17**(4), 565–575 (2007)
6. Czerwiński, R., Kania, D.: Finite State Machine Logic Synthesis for Complex Programmable Logic Devices. Springer, Berlin (2013)
7. Sklyarov, V., Skliarova, I.: Parallel Processing in FPGA-Based Digital Circuits and Systems. TUT Press, Tallinn (2013)
8. Xilinx Corporation Home Page. http://www.xilinx.com

Chapter 5
Hardware Reduction for Mealy UFSMs

Abstract The Chapter deals with optimization of logic circuits of hybrid FPGA-based Mealy FSMs. First of all, the models with two state registers are discussed. This approach allows removal of direct dependence among logical conditions and output functions of Mealy FSM. Next, the proposed design methods are presented. Some improvements are proposed for further hardware reduction. They are based on the special state assignment and transformation of state codes. The proposed methods target joint using such blocks as LUTs, PLAs and EMBs in FSM circuits. The models are discussed based on the principle of object transformation. The last part of the chapter is connected with design methods connected with the object transformation.

5.1 Models with Two State Registers

The hardware reduction for the circuit of BMO is connected with encoding of the collections of microoperations $Y_q \subseteq Y$ [4, 5]. It means that some additional variables are necessary for the encoding. To generate the variables, some resources of FPGA are used. These variables can be eliminated due to using two state registers in Mealy UFSM [3]. Let us discuss this approach.

Let $A_m^h (A_s^h)$ be a conjunction corresponding to the code $K(a_m)$ (the code $K(a_s)$) for the current state (state of transition) from the h-th row of a structure table. Let us represent the term $\dot{F}_h$ corresponding to the h-th row of ST in the following form:

$$\dot{F}_h = A_m^h A_s^h \ (h = 1, \ldots, H_0). \tag{5.1}$$

The conjunction A_m^h includes the internal variables $T_r \in T$. Obviously, the conjunction A_s^h should include some other variables different from $T_r \in T$. Otherwise, all terms (5.1) are equal to zeros.

Let us use the register RG_1 to keep the state codes $K(a_m)$ for current states of UFSM. These states are represented by the state variables $T_r \in T$. Let us use the register RG_2 to keep the state codes $K(a_s)$ for states of transition. Let these states be represented by the additional state variables $\tau_r \in T$. Both registers include R_0 flip-flops. Let us use D flip-flops for implementation both registers RG_1 and RG_2.

© Springer International Publishing AG 2018

A. Barkalov et al., *Logic Synthesis for Finite State Machines Based on Linear Chains of States*, Studies in Systems, Decision and Control 113, DOI 10.1007/978-3-319-59837-6_5

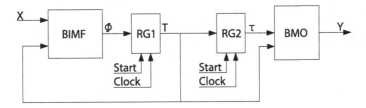

Fig. 5.1 Structural diagram of $P_R Y$ Mealy UFSM

Let us use the symbol $P_R Y$ to denote those Mealy UFSMs. The structural diagram of Mealy UFSM based on (5.1) is shown in Fig. 5.1.

In $P_R Y$ UFSM, the BIMF implements the system (2.9) depending on the terms (2.11). The BMO implements the microoperations $y_n \in Y$ represented by the following system:

$$Y = Y(T, \tau). \tag{5.2}$$

The terms of (5.2) are represented in the form (5.1). The conjunction A_m^h is represented in the form (2.12). The conjunction A_s^h is represented in the following form:

$$A_s^h = \bigwedge_{r_1}^{R_0} \tau_r^{l_{rh}}. \tag{5.3}$$

In (5.3), the symbol l_{rh} stands for the value of the r-th bit of the code $K(a_s)$: $l_{rh} \in \{0, 1\}$, $\tau_r^0 = \bar{\tau}_r$, $\tau_r^1 = \tau_r$ $(r = 1, \ldots, R_0)$. The $P_R Y$ UFSM operates in the following manner. If *Start=1*, then the zero codes are loaded into both registers. In the instant t $(t = 1, 2, \ldots)$ there is a code $K(a_m)$ in RG$_1$ and code $K(a_s)$ in RG$_2$. The BIMF generates functions $D_r \in \Phi$ corresponding to transition number h $(1, \ldots, H_0)$. Using the pulse Clock, the code of the state of transition is loaded into RG$_1$. At the same time the code of the current state is loaded into RG$_2$. Now, the BMO generates the microoperations (5.2). The operation continues till the code $K(a_1)$ is loaded into RG$_1$.

Table 5.1 Structure table for Mealy FSM $P(\Gamma_7)$

a_m	$K(a_m)$	a_s	$K(a_s)$	X_h	Y_h	Φ_h	h
a_1	00	a_2	01	x_1	$y_1 y_2$	D_2	1
		a_3	10	$\bar{x}_1$	y_2	D_1	2
a_2	01	a_2	01	x_2	y_3	D_2	3
		a_4	11	$\bar{x}_2$	$y_1 y_2$	$D_1 D_2$	4
a_3	10	a_4	11	1	$y_3 y_4$	$D_1 D_2$	5
a_4	11	a_1	00	1	y_2	–	6

Let us discuss an example of design for the UFSM $P_R Y(\Gamma_7)$ represented by Table 5.1. As follows from the table, there are $A = \{a_1, \ldots, a_4\}$, $M_0 = 4$, $R_0 = 2$, $X = \{x_1, x_2\}$, $Y = \{y_1, \ldots, y_4\}$, $T = \{T_1, T_2\}$, $\tau = \{\tau_1, \tau_2\}$.

The input memory functions $D_r \in \Phi$ are determined by the following terms: $F_1 = \bar{T}_1 \bar{T}_2 x_1$, $F_2 = \bar{T}_1 \bar{T}_2 \bar{x}_1$, and so on. The terms (5.1) are determined as the following:

$$F_1 = A_2 A_1 = \bar{T}_1 T_2 \bar{\tau}_1 \bar{\tau}_2;$$
$$F_2 = A_3 A_1 = T_1 \bar{T}_2 \bar{\tau}_1 \bar{\tau}_2; \qquad (5.4)$$
$$\vdots \quad \vdots$$
$$F_6 = A_1 A_4 = \bar{T}_1 \bar{T}_2 \tau_1 \tau_2.$$

Now, there is the function $y_1 = \dot{F}_1 \vee \dot{F}_4$. Using terms (5.4), the following Boolean equation can be found

$$y_1 = \bar{T}_1 T_2 \bar{\tau}_1 \bar{\tau}_2 \vee T_1 T_2 \bar{\tau}_1 \tau_2. \qquad (5.5)$$

Similar equations can be obtained for all functions $y_n \in Y$.

This method can be used only if the variables $\dot{F}_h$ $(h = 1, \ldots, H_0)$ are orthogonal. It corresponds to the following conditions:

$$\bigvee_{h=1}^{H_0} \dot{F}_h = 1; \qquad (5.6)$$
$$\dot{F}_i \dot{F}_j = 0 \quad (i \neq j, \ i, j \in \{1, \ldots, H_0\}).$$

If conditions (5.6) are violated, then the behaviour of $P_R Y$ Mealy UFSM differs from the behaviour of the equivalent P Mealy FSM. Let us show it.

Let the following relations take places: $\dot{F}_i = \dot{F}_j$, $Y_i \neq Y_j$. Hear Y_h is a collection of microoperations written in the row number h of ST. Of course, the equality $\dot{F}_i = \dot{F}_j$ is possible if and only if there the equalities $A_m^i = A_m^j$ and $A_s^i = A_s^j$. It means that there are the same current states (states of transitions) for the rows i and j. The output functions $y_n = Y_i \cup Y_j$ are generated if any of transitions i or j is executed. Let us denote as $A(a_m)$ the set of states of transition for the current state of FSM. Let H_m be the number of transitions from the state $a_m \in A_0$. Let us treat the transitions $\langle a_m, a_s \rangle$ as a single transition if the same microoperations are generated for each transition. Let the following condition take place:

$$|A(a_m)| = H_m \quad (m = 1, \ldots, M_0). \qquad (5.7)$$

In this case, a P Mealy FSM can be represented by the model of $P_R Y$ UFSM. If the condition (5.7) is violated, then the laws of behaviour are different for these models.

Let us analyze the structure table of Mealy FSM $P(\Gamma_8)$ represented by Table 5.2. Let ST_m be a subtable of a structure table describing transitions from the state $a_m \in A$.

Table 5.2 Structure table for Mealy FSM $P(\Gamma_8)$

a_m	$K(a_m)$	a_s	$K(a_s)$	X_h	Y_h	Φ_h	h
a_1	000	a_2	001	x_1	$y_1 y_2$	D_3	1
		a_2	001	$\bar{x}_1$	y_3	D_3	2
a_2	001	a_3	010	x_2	$y_4 y_5$	D_2	3
		a_4	011	$\bar{x}_2 x_3$	$y_2 y_3$	$D_2 D_3$	4
		a_2	001	$\bar{x}_2 \bar{x}_3$	y_3	D_3	5
a_3	010	a_4	011	1	$y_2 y_3$	$D_2 D_3$	6
a_4	011	a_5	100	x_2	$y_1 y_5$	D_1	7
		a_5	100	$\bar{x}_2 x_4$	y_3	D_1	8
		a_2	001	$\bar{x}_2 \bar{x}_4$	$y_1 y_2$	D_3	9
a_5	100	a_1	000	1	y_6	–	10

An analysis of Table 5.2 shows that the condition (5.7) is violated for states a_1 and a_4. Let us start from the subtable ST_1. There are the same states of transitions in the rows 1 and 2. So, there is $H_1 = 1$, but is should be equal 2. The microoperations y_1, y_2 and y_3 are generated for the transition number 1, as well as for transition numer 2. There are $H_2 = 3$, $H_3 = 1$ and $H_5 = 1$. But in the case of ST_4, there is $H_4 = 2$. There are three transitions from the state $a_4 \in A$. Therefore, the condition (5.7) is violated. For both transitions from a_4 to a_5, three microoperations are generated (y_1, y_3, y_5). Of course, it does not true for Table 5.2.

To satisfy (5.7), the initial ST should be transformed. The transformation should be executed in such a manner that the condition (5.7) is true for any subtable of the transformed ST. The following approach is proposed for the transformation:

1. Let us analyse the subtable ST_m $(m = 1, \ldots, M_0)$. Let a state a_s appear I times in the column a_s of this subtable. Let us form the set B_s^m corresponding to this situation, where $B_s^m = \{a_s^1, \ldots, a_s^I\}$.
2. Lt us construct the sets $B_s = \bigcup_{m=1}^{M_0}$. These sets corresponds to the states $a_s \in A$.
3. Let us analyze the rows of subtable ST_m $(m = 1, \ldots, M_0)$. Let the state a_s is written in the rows i and j of the ST_m. Let it be $Y_i = Y_j$. In this case, the state a_s is replaced by a single element of B_s. Otherwise, it is replaced by the different elements of B_s.
4. The subtable ST_s $(s = 1, \ldots, M_0)$ is repeated $I_s = |B_s|$ times. A subtable ST_s^i includes all rows from the ST_s but the initial state a_s ir replaced by the state $a_s^i \in B_s$.

In the case of $P(\Gamma_8)$, the following sets can be formed: $B_1 = \{a_1^1\}$, $B_2 = \{a_1^1, a_2^2\}$, $B_3 = \{a_3^1\}$, $B_4 = \{a_4^1\}$ and $B_5 = \{a_5^1, a_5^2\}$. It leads to the following set A_0 for the $P_R Y(\Gamma_8)$: $A_0 = \{a_1^1, a_2^1, a_2^2, a_3^1, a_4^1, a_5^1, a_5^2\}$. Let us encode the states $a_m \in A_0$ by binary codes $K(a_m)$ having R_A bits, where

$$R_A = \lceil \log_2 |A_0| \rceil. \tag{5.8}$$

Table 5.3 Table of Mealy UFSM $P_R Y (\Gamma_8)$

a_m	$K(a_m)$	a_s	$K(a_s)$	X_h	Y_h	Φ_h	h
a_1^1	000	a_2^1	001	x_1	$y_1 y_2$	D_3	1
		a_2^2	010	$\bar{x}_1$	y_3	D_2	2
a_2^1	001	a_3^1	011	x_2	$y_4 y_5$	$D_2 D_3$	3
		a_4^1	100	$\bar{x}_2 x_3$	$y_2 y_3$	D_1	4
		a_2^1	001	$\bar{x}_2 x_3$	y_3	D_3	5
a_2^2	010	a_3^1	011	x_2	$y_4 y_5$	$D_2 D_3$	6
		a_4^1	100	$\bar{x}_2 x_3$	$y_2 y_3$	D_1	7
		a_2^1	001	$\bar{x}_2 \bar{x}_3$	y_3	D_3	8
a_3^1	011	a_4^1	100	1	$y_2 y_3$	D_1	9
a_4^1	100	a_5^1	101	x_2	$y_1 y_5$	$D_1 D_3$	10
		a_5^2	110	$\bar{x}_2 x_4$	y_3	$D_1 D_2$	11
		a_2^1	001	$\bar{x}_2 \bar{x}_4$	$y_1 y_2$	D_3	12
a_5^1	101	a_1^1	000	1	y_6	$-$	13
a_5^2	110	a_1^1	000	1	y_6	$-$	14

In the discussed case, there is $R_A = 3$. Let us encode the states $a_m \in A_0$ in the following way: $K(a_1^1) = 000$, $K(a_1^1) = 001, \ldots, K(a_5^2) = 110$. It leads to the table of Mealy UFSM $P_R Y (\Gamma_8)$ having 14 rows (Table 5.3).

Analysis of Table 5.3 shows that the condition (5.7) takes place for all states $a_m \in A_0$. But the transformed table includes more rows than the initial structure table. In the common case, there are $H(P_R Y)$ rows in the transformed table:

$$H(P_R Y) = \sum_{m=1}^{M_0} H_m I_m = H_0 + \Delta H. \tag{5.9}$$

The value ΔH determines the number of added rows:

$$\Delta H = \sum_{m=1}^{M_0} H_m (I_m - 1). \tag{5.10}$$

If $\Delta H > 0$, then the number of terms in the system (5.4) is increased in comparison with H_0. Also, it can result in the relation

$$R_A > R_0. \tag{5.11}$$

It results in increasing for the hardware amount in the circuit of $P_R Y$ FSM. All these drawbacks can lead to situation when the hardware amount is higher in $P_R Y (\Gamma)$

in comparison with the equivalent Mealy FSM $P(\Gamma)$. Let us discuss the ways leading to deceasing for hardware amount in logic circuits of $P_R Y$ Mealy UFSMs. We consider the case of hybrid FPGAs.

5.2 Optimization of UFSMs with Two Registers

Two different methods can be used for optimization of $P_R Y$ UFSMs:

1. Special state assignment.
2. Transformation of state codes.

Let us consider these methods.

Special state assignment. Analysis of subtables ST_2^1 and ST_2^2 (Table 5.3) shows that the contents of columns $a_s - \Phi_h$ coincide for rows 3 and 6, 4 and 7, and 5 and 8, respectively. The following formulae can be derived from theses subtables:

$$
\begin{aligned}
D_1 &= (A_2^1 \vee A_2^2)\bar{x}_2 x_3; \\
D_2 &= (A_2^1 \vee A_2^2)x_2; \\
D_3 &= (A_2^1 \vee A_2^2)(x_2 \vee \bar{x}_2 \bar{x}_3).
\end{aligned}
\tag{5.12}
$$

All terms of (5.12) include the same part, namely, $A_2^1 \vee A_2^2$.

Let $H(a_s)$ be a set of rows of transformed ST from a subtable ST_s^i ($i = 1, \ldots, I_s$). Using this set, the following formulae can be obtained for input memory functions from subtables ST_s^i:

$$
D_r = \bigvee_{h \in H(a_s)} C_{rh} \Delta_s X_h \quad (r = 1, \ldots, R).
\tag{5.13}
$$

In (5.13), the member Δ_s is determined as $\Delta_s = \bigvee_{i=1}^{I_s} A_s^i$. To optimize functions (5.13), it is necessary to encode the states a_s^i so that the disjunction Δ_s includes the minimum number of terms. For example, let us encode the states a_2^1 and a_2^2 in the following way: $K(a_2^1) = 001$, $K(a_2^2) = 101$. It leads to the equality $\Delta_2 = \bar{T}_2 T_3$. In turn, it results in the following form for the system (5.12):

$$
\begin{aligned}
D_1 &= \bar{T}_2 T_3 \bar{x}_2 x_3; \\
D_2 &= \bar{T}_2 T_3 x_2; \\
D_3 &= \bar{T}_2 T_3 x_2 \vee \bar{T}_2 T_3 \bar{x}_2 x_3.
\end{aligned}
\tag{5.14}
$$

Comparison of (5.12) and (5.14) shows that the number of terms decreases in two times. Let us name this style of state assignment a special state assignment. Let us use the symbol $P_{RO} Y$ for Mealy FSM with the special state assignment. The following method is proposed for synthesis of $P_{RO} Y$ Mealy UFSM:

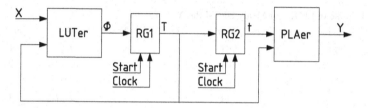

Fig. 5.2 Structural diagram of HFPGA-based $P_{RO}Y$ Mealy UFSM

Fig. 5.3 Outcome of special state assignment for Mealy UFSM $P_{RO}Y(\Gamma_8)$

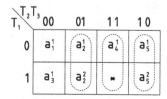

1. Constructing the partition $\Pi_A = \{B_1, \dots, B_M\}$, where $B_m = \{a_m^1, \dots, a_m^{I_m}\}$.
2. Executing the special state assignment.
3. Constructing the transformed structure table of $P_{RO}Y$ Mealy UFSM.
4. Constructing the table of BMO.
5. Deriving functions $\Phi = \Phi(T, X)$ and $Y = (T, \tau)$.
6. Implementing UFSM logic circuit using resources of a particular FPGA chip.

Let us discuss an example of design for the Mealy UFSM $P_{RO}Y(\Gamma_8)$. Let us point out that in the case of hybrid FPGAs, the BIMF is implemented with LUTs and BMO is implemented with PLAs Fig. 5.2.

The following partition Π_A can be found for the Mealy UFSM $P_{RO}Y(\Gamma_8)$: $\Pi_a = \{B_1, \dots, B_5\}$, where $B_1 = \{a_1^1\}$, $B_2 = \{a_2^1, a_2^2\}$, $B_3 = \{a_3^1\}$, $B_4 = \{a_4^1\}$, $B_5 = \{a_5^1 a_5^2\}$. The algorithm JEDI [1] can be used for the special state assignment. One of the variants is shown in Fig. 5.3.

It can be found from the Karnaugh map (Fig. 5.3) that $K(B_1) = 000$, $K(B_2) = *00$, $K(B_3) = 100$, $K(B_4) = *11$ and $K(B_5) = *10$.

To form the table of LUTer (the table of BIMF), it is necessary to form a system of formulae of transitions. In the discussed case, it is the following system:

$$B_1 = x_1 a_2^1 \vee \bar{x}_1 a_2^2;$$
$$B_2 = x_2 a_3^1 \vee \bar{x}_2 x_3 a_4^1 \vee \bar{x}_2 \bar{x}_3 a_2^1;$$
$$B_3 = a_4^1; \qquad\qquad\qquad\qquad (5.15)$$
$$B_4 = x_2 a_5^1 \vee \bar{x}_2 x_4 a_5^2 \vee \bar{x}_2 \bar{x}_4 a_2^1;$$
$$B_5 = a_1^1;$$

There are 10 rows in the table of LUTer (Table 5.4). This number coincides with the number of rows in the initial table of Mealy UFSM (Table 5.2).

Table 5.4 Table of LUTer for Mealy UFSM $P_{RO}Y(\Gamma_8)$

B_m	$K(B_m)$	a_s^i	$K(a_s^i)$	X_h	Y_h	Φ_h	h
B_1	000	a_2^1	001	x_1	$y_1 y_2$	D_3	1
		a_2^2	101	$\bar{x}_1$	y_3	$D_1 D_3$	2
B_2	*01	a_3^1	100	x_2	$y_4 y_5$	D_1	3
		a_4^1	011	$\bar{x}_2 x_3$	$y_2 y_3$	$D_2 D_3$	4
		a_2^1	001	$\bar{x}_2 \bar{x}_3$	y_3	D_3	5
B_3	100	a_4^1	011	1	$y_2 y_3$	$D_2 D_3$	6
B_4	*11	a_5^1	010	x_2	$y_1 y_5$	D_2	7
		a_5^2	110	$\bar{x}_2 x_4$	y_3	$D_1 D_2$	8
		a_2^1	001	$\bar{x}_2 \bar{x}_4$	$y_1 y_2$	D_3	9
B_5	*10	a_1^1	000	1	y_6	–	10

The table of LUTer is used for deriving the system of input memory functions. In the discussed case, the following minimized system Φ can be found:

$$D_1 = \bar{T}_1 \bar{T}_2 \bar{T}_3 \bar{x}_1 \vee \bar{T}_2 T_3 \vee T_2 T_3 \bar{x}_2 x_4;$$
$$D_2 = \bar{T}_2 T_3 \bar{x}_2 x_3 \vee T_1 \bar{T}_2 \bar{T}_3 \vee T_2 T_3 x_2 \vee T_2 T_3 x_4; \qquad (5.16)$$
$$D_3 = \bar{T}_2 \bar{T}_3 \vee \bar{T}_2 T_3 \bar{x}_2 \vee T_2 T_3 \bar{x}_2 \bar{x}_4.$$

The table of BMO (Table of PLAer) has the following columns: $K(a_s)$, $K(B_m)$, X_h, h. It is constructed on the base of the table of LUTer. In the discussed case this table includes 10 rows (Table 5.5). This table is used to construct the system (5.2). For example, the following Boolean equation can be derived from Table 5.5:

$$y_5 = T_1 \bar{T}_2 \bar{T}_3 \bar{\tau}_2 \tau_3 \vee \bar{T}_1 T_2 \bar{T}_3 \tau_2 \tau_3. \qquad (5.17)$$

Let s, t, q be the number of inputs, outputs and product terms of PLA macrocell, respectively. The discussed method has sense if the following condition takes place:

$$S \geq 2R_A. \qquad (5.18)$$

If the condition (5.18) is violated, then the multilevel circuit of PLAer should be implemented [4]. It results in very slow and hardware redundant circuit of the PLAer [3].

Let the following conditions take places:

$$t \geq N; \qquad (5.19)$$
$$q \geq H_{R0}. \qquad (5.20)$$

In (5.20), the symbol H_{R0} stands for the number of rows in the table of LUTer.

Table 5.5 Table of PLAer for Mealy UFSM $P_{RO}Y(\Gamma_8)$

$K(a_s)$	$K(B_m)$	Y_h	h
$T_1 T_2 T_3$	$\tau_1 \tau_2 \tau_3$	$y_1 \ldots y_6$	h
001	000	110000	1
101	000	001000	2
100	*01	000110	3
011	*01	011000	4
001	*01	001000	5
011	100	011000	6
010	*11	100010	7
110	*11	001000	8
001	*11	110000	9
000	*10	000001	10

Fig. 5.4 Implementing PLAer with expansion of outputs

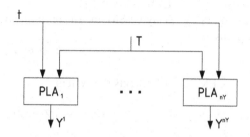

If conditions (5.18)–(5.20) take places, then the block of PLAer is implemented using only a single PLA macrocell. In the discussed case, the following conditions should be true: $S \geq 6$, $t \geq 6$ and $q \geq 10$. If the condition (5.19) is violated, then n_Y PLA macrocells are necessary:

$$n_Y = \langle N/t \rangle. \tag{5.21}$$

In this case, "the expansion of outputs" is executed [4].

It leads to the following circuit (Fig. 5.4).

Each set $Y^j (j = 1, \ldots, n_Y)$ can be viewed as a single block of the partition Π_Y of the set Y. Let it be $t = 3$. In the discussed case, the following partition Π_Y can be found: $\Pi_Y = \{Y^1, Y^2\}$, where $Y^1 = \{y_1, y_2, y_3\}$, $Y^2 = \{y_4, y_5, y_6\}$.

Let the conditions (5.18)–(5.19) take places and let the condition (5.20) be violated. In this case, "the expansion of terms" [4] should be executed. We do not discuss this case in our book. The possible solutions can be found in [4].

Transformation of state codes. It is quite possible a situation when the special state assignment cannot decrease the number of rows in the table of $P_R Y$ UFSM. In this case, we propose to introduce a special code transformer. It executes the transformation of state codes for states $a_m^i \in B_m$ into class codes. It allows providing the following features:

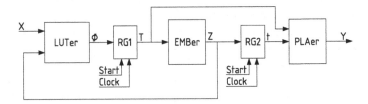

Fig. 5.5 Structural diagram of $\text{P}_{\text{RE}}Y$ Mealy UFSM

1. Decreasing the number of terms in the system Φ up to H_0, as well as the number of terms in the system Y.
2. Decreasing the number of inputs of LUTer up to $L + R_0$.

Let us encode each class $B_m \in \Pi_A$ by the binary code $K(B_m)$ having R_0 bits. Let us use the variables $z_r \in Z$ for the class encoding, where $|Z| = R_0$. It leads to the $\text{P}_{\text{RE}}Y$ Mealy UFSM (Fig. 5.5).

In $\text{P}_{\text{RE}}Y$ Mealy UFSM, the LUTer implements the system of input memory functions

$$\Phi = \Phi(Z, X). \tag{5.22}$$

The EMBer implements the system

$$Z = Z(T). \tag{5.23}$$

The system (5.23) can be implemented using a single EMB if the following condition takes place:

$$V_0 \geq R_0 \cdot 2^{R_A}. \tag{5.24}$$

The $\text{P}_{\text{RE}}Y$ UFSM operates in the following manner. In the beginning of each cycle, the RG1 contains a code of the current state $a_m^i \in B_m$. The EMBer transforms this code into the code $K(B_m)$. Using the pulse Clock, this code is loaded into RG2. At the same time, the code $K(a_s^i)$ is loaded into RG1, where $a_s^i \in A$ is a state of transition. The PLAer generates the microoperations $y_n \in Y$ represented by the system (5.2). The operation is continued till the code $K(a_1)$ is loaded into RG1.

There are the following steps in the proposed method of synthesis for $\text{P}_{\text{RE}}Y$ Mealy UFSM:

1. Constructing the table of transitions for $\text{P}_R Y$ Mealy UFSM and finding the partition Π_A.
2. Encoding the classes $B_m \in \Pi_a$ using variables $z_r \in Z$.
3. Constructing the table of LUTer.
4. Constructing the table of EMBer.
5. Constructing the table of PLAer.
6. Implementing the logic circuit of $\text{P}_{\text{RE}}Y$ Mealy UFSM.

Table 5.6 Table of LUTer for Mealy UFSM $P_{RE}Y(\Gamma_8)$

B_m	$K(B_m)$	a_s^i	$K(a_s^i)$	X_h	Y_h	Φ_h	h
B_1	000	a_2^1	001	x_1	$y_1 y_2$	D_3	1
		a_2^2	010	$\bar{x}_1$	y_3	D_2	2
B_2	*01	a_3^1	011	x_2	$y_4 y_5$	$D_2 D_3$	3
		a_4^1	100	$\bar{x}_2 x_3$	$y_2 y_3$	D_1	4
		a_2^1	001	$\bar{x}_2 \bar{x}_3$	y_3	D_3	5
B_3	*01	a_4^1	100	1	$y_2 y_3$	D_1	6
B_4	*11	a_5^1	101	x_2	$y_1 y_5$	$D_1 D_3$	7
		a_5^2	110	$\bar{x}_2 x_4$	y_3	$D_1 D_2$	8
		a_2^1	001	$\bar{x}_2 \bar{x}_4$	$y_1 y_2$	D_3	9
B_5	1**	a_1^1	000	1	y_6	–	10

Let us discuss an example of desing for Mealy UFSM $P_{RE}Y(\Gamma_8)$. The partition Π_A is constructed before. It includes the classes $B_1 = \{a_1^1\}$, $B_2 = \{a_2^1, a_2^2\}$, $B_3 = \{a_3^1\}$, $B_4 = \{a_4^1\}$ and $B_5 = \{a_5^1, a_5^2\}$. In the discussed case, there is $R_0 = 3$, therefore, there is the set $Z = \{z_1, z_2, z_3\}$. Let us encode the classes in the following way: $K(B_1) = 000, \ldots, K(B_5) = 100$. Let the states $a_m^i \in A$ be encoded in the same way as it is shown in Table 5.3.

The table of LUTer (Table 5.6) is constructed on the base of the system (5.15). It is similar to Table 5.4.

Let us point out that some positions in the codes $K(B_m)$ include the signs "*". It is obtained taking into account the unused input assignments of variables $z_r \in Z$. There are three unused codes: 101, 110 and 111.

The table of LUTer is used for constructing the system (5.22). For example, the following Boolean function can be derived from Table 5.6:

$$D_2 = \bar{z}_1 \bar{z}_2 \bar{z}_3 \bar{x}_1 \vee \bar{z}_2 z_3 x_2 \vee z_2 z_3 \bar{x}_2 x_4. \tag{5.25}$$

The table of EMBer includes the columns a_m^i, B_m, $K(a_m^i)$, $K(B_m)$, m. The column $K(B_m)$ contains an address of the cell, whereas the column $K(B_m)$ determines its content. It is Table 5.7 in the case of Mealy UFSM $P_{RE}Y(\Gamma_8)$.

Of course, it is possible to use LUTs for implementing the system (5.23). In the discussed case, the following functions can be obtained:

$$z_1 = T_1 T_2 \vee T_1 T_2;$$
$$z_2 = T_2 T_3 \vee T_1 \bar{T}_2 \bar{T}_3; \tag{5.26}$$
$$z_3 = \bar{T}_1 \bar{T}_2 T_3 \vee \bar{T}_1 T_2 \bar{T}_3.$$

The table of PLAer is practically the same for Mealy UFSM $P_{RO}Y(\Gamma)$ and $P_{RE}Y(\Gamma)$. We do not show this table for the Mealy UFSM $P_{RE}Y(\Gamma_8)$. Now, let us

Table 5.7 Table of EMBer for Mealy UFSM $P_{RE}Y(\Gamma_8)$

a_m^i	B_m	$K(a_m^i)$	$K(B_m)$	m
		$T_1T_2T_3$	$z_1z_2z_3$	
a_1^1	B_1	000	000	1
a_2^1	B_2	001	001	2
a_2^2	B_2	010	001	3
a_3^1	B_3	011	010	4
a_4^1	B_4	100	011	5
a_5^1	B_5	101	100	6
a_5^2	B_5	110	100	7

Fig. 5.6 Outcome of special state assignment

$T_1 \backslash T_2T_3$	00	01	11	10
0	a_1^1	a_2^1	a_2^2	a_2^3
1	a_3^1	a_4^1	a_4^2	a_4^3

discuss the following example. Let it be the following partition $\Pi_A = \{B_1, \ldots, B_4\}$, where $B_1 = \{a_1^1\}$, $B_2 = \{a_2^1, a_2^2, a_2^3\}$, $B_3 = \{a_3^1\}$, $B_4 = \{a_4^1, a_4^2, a_4^3\}$. Let it be $H_1 = 2$, $H_2 = 6$, $H_3 = 3$ and $H_4 = 7$. The transformed table of P_RY UFSM includes 44 rows. (It can be calculated using (5.8)). Hence, the system (5.2) includes 44 terms.

Let us use the special state assignment (Fig. 5.6).

As follows from Fig. 5.6, the class B_2 is represented by the codes 0*1 and 01*, whereas the class B_4 by the codes 11* and 1*1. So, the table of LUTer includes 31 rows for the case of P_{RO}Y Mealy UFSM.

Because of $I = 4$, there is $R_A = 2$. It means that the condition (5.11) takes place. If the model of P_{RE}Y Mealy UFSM is used, then the table of LUTer contains only $H_0 = 18$ rows. At the same time, there is a decreasing for the number of inputs of LUTer. So, the code transformation is the most efficient if the condition (5.11) takes place.

5.3 Principle of Object Code Transformation

As it was mentioned before, the hardware reduction for FSM logic circuit is connected with the structural decomposition, which in turn is connected with increase for the number of levels in the FSM model. To optimize the hardware amount in block BMO, it is necessary to generate some additional variables for encoding of microoperations (or collections of microoperations). The methods discussed in this Chapter are taken from [2]. These methods are based on one-to-one match among

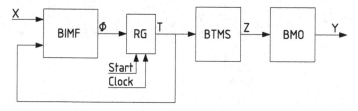

Fig. 5.7 Structural diagram of Mealy FSM1

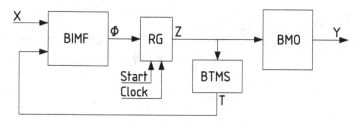

Fig. 5.8 Structural diagram of Mealy FSM2

collections of microoperations and states. There are two objects of FSM, namely, its internal states $a_m \in A$ and collections of microoperations $Y_t \subseteq Y$. Let us point out that states and collections of microoperations are heterogeneous objects respectively each other, whereas different states, for example, are the homogenous objects respectively each other. The optimization methods discussed in this Chapter are based on identification of one-to-one match among heterogeneous objects. If this match is found, then the block BIMF generates only codes for one object (which is a primary object), while a special code transformer generates the codes of another object (which is a secondary object). Let us name these approaches as the methods of object code transformation (OCT).

Let us find a one-to-one match $A \rightarrow Y$ among the states as primary objects and the microoperations as secondary objects. In this case, the block BIMF generates the state variables $T_r \in T = \{T_1, \ldots, T_R\}$ to encode the states, whereas a special state code transformer block BTSM generates variables $z_r \in Z$ used for encoding of collections of microoperations. The structural diagram of Mealy FSM based on this principle is shown in Fig. 5.7. Let the symbol $PC_A Y$ stand for this model if collections of microoperations are encoded, whereas the symbol $PC_A D$ stands for encoding of the classes of compatible microoperations. Let us name such models as FSM1.

Let us find a one-to-one match $Y \rightarrow A$ among the microoperations as primary objects and the states as secondary objects. In this case, the block BIMF generates variables $z_r \in Z$, whereas a special microoperation code transformer block BTMS generates state variables $T_r \in T$. This approach results in the models of FSM2, denoted as $PC_Y Y$ (if collections of microoperations are encoded) or as $PC_Y D$ (if classes of compatible microoperations are encoded). Their structural diagram is shown in Fig. 5.8.

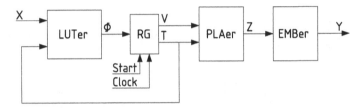

Fig. 5.9 FPGA-based model of Mealy FSM1

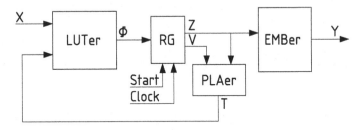

Fig. 5.10 FPGA-based model of Mealy FSM2

These models correspond to cases when an FSM has the same numbers of states and collections of microoperations. If this condition is violated, then some additional identifiers should be used belonging to a set of identifiers V. In common case, the block BIMF generates variables T and V (Fig. 5.9) or variables Z and V (Fig. 5.10). All these variables are the outputs of the register RG.

When FPGAs are used for implementing these FSMs, the BIMF is implemented as LUTer, the BMO as EMBer. Blocks BTSM and BTMS can be implemented using either EMBs or PLAs. The second case is represented by the structural diagrams shown in Figs. 5.9 and 5.10.

Of course, the EMBer can be absent. It is possible in the case of unitary encoding of microoperations. In this case, microoperations $y_n \in Y$ are the output of RG, as well as the additional variables $v_r \in V$.

Thus, in common case the number of bits in the register RG for Mealy FSM with object code transformation exceeds this number for equivalent PY or PD Mealy FSM. Obviously, the proposed approach can be applied only when the total hardware amount for blocks BIMF and BTSM (BTMS) is less, than the hardware amount for block BIMF of PY (PD) Mealy FSM. The same approach can be applied for Moore FSM but it is out the scope of this book.

5.4 Design of Mealy FSM1 with OCT

Let the Mealy FSM $P(\Gamma_9)$ be specified by its structure table (Table 5.8). Consider logic synthesis for models of PC_AY, PC_AD, PC_YY, and PC_YD Mealy FSM, based on the Mealy FSM

The following procedure is proposed for logic synthesis of Mealy FSM1:

1. **One-to-one identification of collections of microoperations.** Let $Y(a_s)$ be a set of collections of microoperations generated under transitions into the state $a_s \in A$, where $n_s = |T(a_s)|$. In this case, it is necessary n_s identifiers for one-to-one identifications of collections $Y_t \subset Y(a_s)$. In common case, it is enough $K = max(n_1, \ldots, n_M)$ identifiers for one-to-one identification of all collections $Y_t \subset Y$, these identifiers form the set $I = \{I_1, \ldots, I_K\}$. Let us encode each identifier $I_K \in I$ by a binary code $K(I_K)$ having $R_V = \lceil \log_2 K \rceil$ bits. Let us use the variables $v_r \in V = \{v_1, \ldots, v_{RV}\}$ for encoding of the identifiers.

   Let each collection $Y_t \in (a_s)$ correspond to the pair $\beta = \langle I_k, a_s \rangle$ where $I_k \in I$. Of course, an identifier $I_k \in I$ should be different for different collections. In this case, a code $K(I_k)$ of the set $Y_t \in Y(a_s)$ is determined by the following concatenation:

$$K(Y_t) = K(I_k) * K(a_s). \tag{5.27}$$

   In (5.27) the symbol * stands for the concatenation of these codes.

2. **Encoding of collections of microoperations.** If the method of maximal encoding of collections of microoperations is applied, then let a collection $Y_t \subset Y$ be determined by a binary code $C(Y_t)$ having $Q = \lceil \log_2 T_0 \rceil$ bits, where T_0 is the number of collections. If the method of encoding of the classes of compatible

Table 5.8 Structure table of Mealy FSM 9

a_m	$K(a_m)$	a_s	$K(a_s)$	X_h	Y_h	Φ_h	h
a_1	000	a_2	010	x_1	$y_1 y_2$	$D_1 D_2$	1
		a_3	011	$\bar{x}_1$	y_3	D_3	2
a_2	010	a_2	010	x_2	$y_1 y_2$	D_2	3
		a_3	011	$\bar{x}_2 x_3$	y_4	D_2	4
		a_4	100	$\bar{x}_2 \bar{x}_3$	$y_1 y_2$	$D_1 D_3$	5
a_3	011	a_4	100	x_1	$y_2 y_5$	D_1	6
		a_5	101	$\bar{x}_1$	y_6	$D_1 D_3$	7
a_4	100	a_5	101	1	$y_3 y_7$	$D_1 D_3$	8
a_5	101	a_2	010	$x_2 x_3$	$y_1 y_2$	D_2	9
		a_3	011	$x_2 \bar{x}_3$	y_3	$D_2 D_3$	10
		a_5	101	$\bar{x}_2 x_4$	$y_3 y_7$	$D_1 D_3$	11
		a_1	000	$\bar{x}_2 \bar{x}_4$	$-$	$-$	12

microoperations is used, then any collection Y_t is represented as the following vector:

$$Y_t = \langle y_t^1, y_t^2, \ldots, y_t^J \rangle. \tag{5.28}$$

In (5.28), the symbol J stands for the number of the classes of compatible microoperations, whereas the symbol y_t^j denotes microoperation $y_n \in Y_t$, belonged to the class j of compatible microoperations ($j = 1, \ldots, J$). Therefore, a code of collection of mircrooperations Y_t is represented as a concatenation of microoperation codes.

3. **Construction of transformed structure table.** For FSM1, the transformed ST is used to generate input memory functions Φ and additional functions of identification V. These systems depend on the same terms and are represented as the following:

$$\phi_r = \bigvee_{h=1}^{H} C_{rh} A_m^h X_h \quad (r = 1, \ldots, R), \tag{5.29}$$

$$v_r = \bigvee_{h=1}^{H} C_{rh} A_m^h X_h \quad (r = 1, \ldots, R_V). \tag{5.30}$$

Obviously, to represent system (5.30) the column Y_h of initial ST should be replaced by columns: I_h is an identifier of the collection Y_h from pair $\beta_{s,h}$; $K(I_h)$ is a code of identifier I_k; V_h are variables $v_r \in V$, equal to 1 in the code $K(I_h)$.

4. **Specification of block BTSM.** The block BTSM generates variables $z_q \in Z$ represented as the following functions

$$Z = Z(V, T). \tag{5.31}$$

To construct system (5.31), it is necessary to built a table with columns a_s, $K(a_s)$, I_k, $K(I_k)$, Y_h, Z_h, h. The table includes all pairs $\beta_{t,s}$, determined the collection Y_1, next all pairs determined the collection Y_2, and so on. The number of their rows (H_0) is determined as a result of summation for numbers n_s ($S = 1, \ldots, M$). The column of the table includes variables $z_q \in Z$, equal to 1 in the code $K(Y_h)$. The system (5.31) can be represented as the following:

$$z_q = \bigvee_{h=1}^{H_0} C_{qh} X_h A_s^h \quad (q = 1, \ldots, Q). \tag{5.32}$$

In (5.32) the symbol V_h stands for conjunction of variables $v_r \in V$, corresponded to the code $K(I_k)$ of identifier from the row h of this table.

5. **Specification of block for generation of microoperations.** This step is executed in the same manner, as it is done for PY or PD Mealy FSMs.

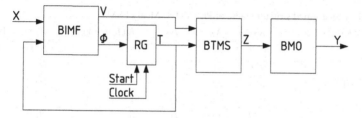

Fig. 5.11 Structural diagram of PC_AY Mealy FSM

6. **Synthesis of FSM logic circuit.** For the Mealy FSM1, there is no need in keeping codes of identifiers in the register RG. Therefore, these FSM models should be refined. The structural diagram of $PC_A Y$ Mealy FSM is shown in Fig. 5.11. It is the same as the structural diagram $PC_A D$ Mealy FSM. In both cases, the block BIMF implements functions (5.29)–(5.30), the block BTSM generates functions (5.31), block BMO implements microoperations $Y = Y(Z)$.

In Table 5.8 there are the following collections of microoperations $Y_1 = \emptyset$, $Y_2 = \{y_1, y_2\}$, $Y_3 = \{y_3\}$, $Y_4 = \{y_4\}$, $Y_5 = \{t_5\}$, $Y_6 = \{Y_6\}$, $Y_7 = \{y_7\}$, $T_0 = 7$.

Let us consider an example of logic synthesis for the $PC_A Y(\Gamma_9)$ Mealy FSM. The following sets can be derived from Table 5.8: $Y(a_1) = \{Y_1\}$, $Y(a_2) = \{Y_2\}$, $Y(a_3) = \{Y_3, Y_4\}$, $Y(a_4) = \{Y_2, Y_5\}$, $Y(a_5) = \{Y_6, Y_7\}$. It gives the value $K = 2$. Thus, it is enough two identifiers creating the set $I = \{I_1, I_2\}$; they can be encoded using $R_V = 1$ variables from the set $V = \{v_1\}$. Let $K(I_1) = 0$, $K(I_2) = 1$, then the following codes can be obtained using formula (5.27): $K(Y_1) = *000$, $K(Y_2^1) = *010$, $K(Y_2^2) = 0100$, $K(Y_3) = 0011$, $K(Y_4) = 1011$, $K(Y_5) = 1100$, $K(Y_6) = 0101$ and $K(Y_7) = 1101$. This example shows that there are m_t different codes determined a collection $Y_t \subset Y$ if this collection belongs to m_t different sets $Y(a_s)$. For example, for the collection $Y_2 \in Y(a_2) \cap Y(a_1)$ we have $m_2 = 2$, thus the collection Y_2 corresponds to codes $K(Y_2^1)$ and $K(Y_2^2)$.

There are $T_0 = 7$ different collections, thus $R_Y = 3$ and $Z = \{z_1, z_2, z_3\}$. Let the collections $Y_t \subset Y$ be encoded in the following way: $K(Y_1) = 000$, $K(Y_2) = 001, \ldots, K(Y_7) = 110$. The transformed structure table (Table 5.9) should be constructed to find functions (5.29)–(5.30).

If the condition $n_s = 1$ takes place for some collection $Y_t \in Y(a_s)$, then there is no need in identifier code for this collection. This situation is marked by the symbol "-" in the corresponding row of transformed structure table. As it was mentioned, we can derive systems (5.29)–(5.30) from Table 5.9. For example, the following SOPs can be found: $v_1 = F_4 \vee F_5 \vee F_8 \vee F_{11} = A_2\bar{x}_2x_3 \vee A_3x_4 \vee \ldots = \bar{T}_1T_2\bar{T}_3\bar{x}_2x_3 \vee \bar{T}_1T_2T_3x_4 \vee \ldots$, $D_2 = F_2 \vee F_3 \vee F_4 \vee F_9 \vee F_{10} = A_1\bar{x}_1 \vee A_2x_2 \vee \ldots = \bar{T}_2\bar{T}_3x_2 \vee \bar{T}_1T_2\bar{T}_3x_2 \ldots$. Both systems are irregular, thus they are implemented using LUTs.

Table 5.10 specifies the block BTMS; it includes $H_0 = 8$ rows. This number is equal to the outcome of summation for the numbers $n_s(S = 1, \ldots, 5)$. System is irregular and it is implemented using PLAs. For example, the SOP $z_1 = F_6 \vee F_7 \vee$

Table 5.9 Transformed structure table of $PC_A Y(\Gamma_9)$ Mealy FSM

a_m	$K(a_m)$	a_s	$K(a_s)$	X_h	I_h	$K(I_k)$	V_h	Φ_h	h
a_1	000	a_2	010	x_1	–	–	–	$D_1 D_2$	1
		a_3	011	$\bar{x}_1$	I_1	0	–	D_3	2
a_2	010	a_2	010	x_2	–	–	–	D_2	3
		a_3	011	$\bar{x}_2 x_3$	I_2	1	v_1	D_2	4
		a_4	100	$\bar{x}_2 \bar{x}_3$	I_1	0	–	$D_1 D_3$	5
a_3	011	a_4	100	x_1	I_2	1	v_1	D_1	6
		a_5	101	$\bar{x}_1$	I_1	0	–	$D_1 D_3$	7
a_4	100	a_5	101	1	I_2	1	v_1	$D_1 D_3$	8
a_5	101	a_2	010	$x_2 x_3$	–	–	–	D_2	9
		a_3	011	$x_2 \bar{x}_3$	I_1	0	–	$D_2 D_3$	10
		a_5	101	$\bar{x}_2 x_4$	I_2	1	v_1	$D_1 D_3$	11
		a_1	000	$\bar{x}_2 \bar{x}_4$	–	–	–	–	12

Table 5.10 Specification of block BTSM of Mealy $PC_A Y(\Gamma_9)$ FSM

a_s	$K(a_s)$	I_h	$K(I_k)$	Y_h	Z_h	h
a_1	000	–	–	Y_1	–	1
a_2	010	–	–	Y_2	z_3	2
a_3	011	I_1	0	Y_3	z_2	3
a_3	011	I_2	1	Y_4	$z_2 z_3$	4
a_4	100	I_1	0	Y_2	z_3	5
a_4	100	I_2	1	Y_5	z_1	6
a_5	101	I_1	0	Y_6	$z_1 z_3$	7
a_5	101	I_2	1	Y_7	$z_1 z_2$	8

$F_8 = A_4 v_1 \vee A_5 \bar{v}_1 = T_1 \bar{T}_2 \bar{T}_3 v_1 \vee T_1 \bar{T}_2 T_3 \bar{v}_1 \vee T_1 T_2 T_3 \bar{v}_1 \vee T_1 \bar{T}_2 T_3 \vee v_1$ can be derived from the table specified the block BTSM in our example.

The block BMO is specified by the table of microoperations. For the $PC_A Y(\Gamma_9)$ Mealy FSM, this table includes $T_0 = 8$ rows (Table 5.11). Let us point out that codes $C(Y_t)$ are used as codes of collections Y_t.

Let us consider an example of the logic synthesis for the Mealy FSM $PC_A D(\Gamma_9)$. Obviously, the outcome of one-to-one identification is the same for equivalent $PC_A Y$ and $PC_A D$ Mealy FSMs. To encode the collections of microoperations, it is necessary to find the partition Π_Y of the set of microoperations Y by the classes of pseudoequivalent microoperations. For the discussed example, the following partition $\Pi_Y = \{Y^1, Y^2\}$ with two classes can be found, where $Y^1 = \{y_1, y_3, y_4, y_5\}$, $Y_2 = \{y_2, y_6, y_7\}$. It is enough $Q_1 = 3$ variables to encode the microoperations $y_n \in Y_1$, and $Q_2 = 2$ variables for the microoperations $y_n \in Y_2$. It means that there is the set $Z = \{z_1, \ldots, z_5\}$, its cardinality is found as $Q = Q_1 + Q_2 = 5$. Let us encode microoperations $y_n \in Y$ in the way shown in Table 5.12. It leads to the

Table 5.11 Specification of microoperations of $PC_A Y(\Gamma_9)$ Mealy FSM

Y_t	$C(Y_t)$	$y_1 y_2 \ldots y_7$
Y_1	000	0000000
Y_2	001	1100000
Y_3	010	0010000
Y_4	011	0001000
Y_5	100	0100100
Y_6	101	0000010
Y_7	110	0010001

Table 5.12 Codes of microoperations for Mealy FSM $PC_A D(\Gamma_9)$

Y^1	$K(Y_n^1)$	Y^2	$K(Y_n^2)$
	$z_1 z_2 z_3$		$z_4 z_5$
y_1	001	y_2	01
y_3	010	y_6	10
y_4	011	y_7	11
y_5	100	–	–

Table 5.13 Codes of collections of microoperations for Mealy FSM $PC_A D(\Gamma_9)$

t	Y^1	$C(Y_t)$	t	Y^2	$C(Y_t)$
1	$\varnothing$	00000	5	$y_2 y_5$	10001
2	$y_1 y_2$	00101	6	y_6	00010
3	y_3	01000	7	$y_3 y_7$	01011
4	y_4	01100			

Table 5.14 Specification of block BTSM Mealy FSM $PC_A D(\Gamma_9)$

a_s	$K(a_s)$	I_h	$K(I_k)$	Y_h	Z_h	h
a_1	000	–	–	Y_1	–	1
a_2	010	–	–	Y_2	$z_3 z_5$	2
a_3	011	I_1	0	Y_3	z_2	3
a_3	011	I_2	1	Y_4	$z_2 z_3$	4
a_4	100	I_1	0	Y_2	$z_3 z_5$	5
a_4	100	I_2	1	Y_5	$z_1 z_5$	6
a_5	101	I_1	0	Y_6	$z_1 z_4$	7
a_5	101	I_2	1	Y_7	$z_1 z_4 z_5$	8

codes $C(Y_t)$ of collections $Y_t \in Y$ shown in Table 5.13. Let us point out that if some microoperation $y_n^j \notin Y_t$, then the field j of code $C(Y_t)$ contains only zeros.

The transformed structure table of Mealy FSM $PC_A D(\Gamma_9)$ is identical to Table 5.9. The table specifying the block BTSM for both models is constructed in the same way. As a rule, this table for $PC_A D$ Mealy FSM includes more variables $z_r \in Z$ (Table 5.14 in our example), than its counterpart for $PC_A Y$ Mealy FSM.

There is no need in a table specifying microoperations, because Table 5.12 contains inputs and outputs for decoders of the block BMO.

5.5 Design of Mealy FSM2 with OCT

The following procedure is proposed to design a Mealy FSM2:

1. **One-to-one identification of states.** Let $A(Y_t)$ be a set of states, such that a collection $Y_t \subset Y$ is generated under some transitions in these states, and let $m_t = |A(Y_t)|$. In this case, it is enough m_t identifiers for one-to-one identifications of the states $a_m \in A(Y_t)$. It is necessary $K = \max(m_1, \ldots, m_T)$ variables for one-to-one identification of the states $a_m \in A$. Let these identifiers form a set I. Let us encode an identifier $I_k \in I$ by a binary code $K(I_k)$ and let us construct a set of variables $V = \{v_1, \ldots, v_{R1}\}$ used for encoding of identifiers, where $R_l = \lceil \log_2 K \rceil$. Let each state $a_s \in A(Y_t)$ correspond to a pair $\alpha_{t,s} = \langle I_k, Y_t \rangle$, then the code for state a_s is determined by the following concatenation:

$$C(a_s) = K(Y_t) * K(I_k) \qquad (5.33)$$

2. **Encoding of collections of microoperations.** This step is executed using the approach discussed before.

3. **Construction of transformed structure table.** This table is used to derive functions $V = V(T, X)$ and $Z = Z(T, X)$. To construct it, the columns a_s, $K(a_s)$, Φ_h are eliminated from the initial structure table. At the same time the column Y_h is replaced by columns V_h and Z_h. The column Z_h contains variables $z_q \in Z$ equal to 1 in the code $K(Y_h)$. The system Z includes the following equations:

$$z_q = \bigvee_{h=1}^{H} C_{qh} A_m^h X_h \quad (q = 1, \ldots, Q). \qquad (5.34)$$

4. **Specification of code transformer.** The code transformer BTMS generates functions

$$\Phi = \Phi(V, Z). \qquad (5.35)$$

This system can be specified by a table with the following columns: Y_t, $K(Y_t)$, I_k, $K(I_k)$, a_s, $K(a_s)$, Φ_h, h. The table includes all pairs $\angle I_k, Y_t \rangle$ for the state a_1, next, all pairs for a_2, and so on. The number of rows H_0 in this table is determined

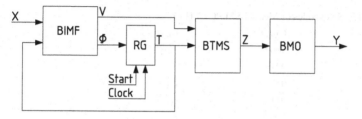

Fig. 5.12 Structural diagram of $PC_Y Y$ and $PC_A D$ Mealy FSMs

as a result of summation for the numbers $m_t (t = 1, \ldots, T)$. The system of input memory functions is represented as the following one:

$$\phi_r = \bigvee_{h=1}^{H_0} C_{rh} V_h Z_h \quad (r = 1, \ldots, R). \tag{5.36}$$

In (5.36) the symbol Z_h stands for conjunction of variables $z_r \in Z$ corresponded to the collection of microoperations $Y_t \subset Y$ from the row h of the table specifying block BTMS.

5. **Construction of the table of microoperations.** This step is executed using the same approach as the one applied for $PC_A Y$ Mealy FSM.

6. **Synthesis of FSM logic circuit.** For structural diagram shown in Fig. 5.8 the number of bits in the register RG is equal to $Q + R_V$. This number can be decreased up to R, using the structural diagrams shown in Fig. 5.12. In this case, the block BTMS generates input memory functions instead of state variables T. Due to such approach, it is enough R flip-flops in the register RG.

Let us discus an example of logic synthesis for the Mealy FSM $PC_Y Y (\Gamma_9)$. For FSM 9 there are $T_0 = 7$ collections of microoperations, namely: $Y_1 = \emptyset$, $Y_2 = \{y_1, y_2\}$, $Y_3 = \{y_3\}$, $Y_4 = \{y_4\}$, $Y_5 = \{y_2, y_5\}$, $Y_6 = \{y_6\}$, $Y_7 = \{y_3, y_7\}$ (Table 5.8). Let us construct the sets $A(Y_t)$ and define their cardinality numbers: $A(Y_1) = \{a_1\}$, $m_1 = 1$; $A(Y_2) = \{a_2, a_4\}$, $m_2 = 2$; $A(Y_3) = \{a_3\}$, $m_3 = 1$; $A(Y_4) = \{a_3\}$, $m_4 = 1$; $A(Y_5) = \{a_4\}$, $m_5 = 1$; $A(Y_6) = \{a_5\}$, $m_6 = 1$, and $A(Y_7) = \{a_5\}$, $m_7 = 1$. Thus, it is enough $K = 2$ identifiers, that is $I = \{I_1, I_2\}$. The identifiers $I_k \in I$ can be encoded using $R_V = 1$ variable, that is $V = \{v_1\}$. Let the identifiers be encoded in the following way: $K(I_1) = 0$ and $K(I_2) = 1$. Let us find the pairs $\alpha_{t,s}$ for each element from the sets $A(Y_t)$. If $m_t = 1$, then the first component of corresponding pair is represented by the symbol $\emptyset$. This symbol corresponds to uncertainty in the code $C(a_s)^t$, where the superscript t means that the code of state a_s belongs to the pair $\alpha_{t,s}$. The following pairs can be constructed in the discussed example: $\alpha_{1,1} = \langle \emptyset, Y_1 \rangle$, $\alpha_{2,2} = \langle I_1, Y_2 \rangle$, $\alpha_{2,4} = \langle I_2, Y_2 \rangle$, $\alpha_{3,3} = \langle \emptyset, Y_3 \rangle$, $\alpha_{4,3} = \langle \emptyset, Y_4 \rangle$, $\alpha_{5,4} = \langle \emptyset, Y_5 \rangle$, $\alpha_{6,5} = \langle \emptyset, Y_6 \rangle$, $\alpha_{7,5} = \langle \emptyset, Y_7 \rangle$. Using these pairs together with (5.8), we can get the codes $C(a_s)$ shown in Table 5.15.

Table 5.15 State codes of Mealy FSM $PC_Y Y (\Gamma_9)$

a_m	$C(a_s)^t$	α_{t_m}	h
a_1	000*	$\alpha_{1,1}$	1
a_2	0010	$\alpha_{2,2}$	2
a_3	010*	$\alpha_{3,3}$	3
a_3	011*	$\alpha_{4,3}$	4
a_4	100*	$\alpha_{5,4}$	5
a_4	0011	$\alpha_{2,4}$	6
a_5	101*	$\alpha_{6,5}$	7
a_5	110*	$\alpha_{7,5}$	8

Table 5.16 Transformed structure table of Mealy FSM $PC_Y Y (\Gamma_9)$

a_m	$K(a_m)$	X_h	Z_h	V_h	h
a_1	000	x_1	z_3	–	1
		$\bar{x}_1$	z_2	–	2
a_2	010	x_2	z_3	–	3
		$\bar{x}_2 x_3$	$z_2 z_3$	–	4
		$\bar{x}_2 \bar{x}_3$	z_3	v_1	5
a_3	011	x_1	z_1	–	6
		$\bar{x}_1$	$z_1 z_3$	–	7
a_4	100	1	$z_1 z_2$	–	8
a_5	101	$x_2 x_3$	z_3	–	9
		$x_2 \bar{x}_3$	z_2	–	10
		$\bar{x}_2 x_4$	$z_1 z_2$	–	11
		$\bar{x}_2 \bar{x}_4$	–	–	12

This table includes $H_0 = m_1 + \ldots + m_T$ rows. As follows from Table 5.15 each from the states a_3, a_4, and a_5 have two different codes of the type (5.33). In the common case, the number of codes $C(a_s)^t$ for some state $a_m \in A$ is equal to the number of different sets $A(Y_t)$, including this state a_m. The codes of collections of microoperations shown in Table 5.15 are the same as they were obtained before. The codes are placed in the three most significant positions of the column $C(a_m)$.

Using the known method, we can construct the transformed structure table of Mealy FSM (Table 5.16) on the base of the initial structure table (Table 5.8). Using Table 5.16, we can derive systems (5.34) and (5.30).

The table used for specification of the block $BTMS$ (Table 5.17) includes $H_2 = 2^{R_0} - H_1$ rows, where $R_0 = \lceil \log_2 H_0 \rceil$. It is necessary if the logic circuit of BTMS is implemented with embedded memory blocks. In this case all possible addresses should be present. Let us point out that at least $H_1 = (2^Q - T) 2^{R_1}$ rows contain zero output codes corresponded to unused collections of microoperations. For the FSM

Table 5.17 Specification of block BTMS for Mealy FSM $PC_Y Y(\Gamma_9)$

Y_t	$K(Y_T)$	I_k	$K(I_k)$	a_s	$K(a_s)$	Φ_h	h
Y_1	000	–	0	a_1	000	–	1
	000	–	1	a_1	000	–	2
Y_2	001	I_1	0	a_2	010	D_2	3
	001	I_2	1	a_4	100	D_1	4
Y_3	010	–	0	a_3	011	$D_2 D_3$	5
	010	–	1	a_3	011	$D_2 D_3$	6
Y_4	011	–	0	a_3	011	$D_2 D_3$	7
	011	–	1	a_3	011	$D_2 D_3$	8
Y_5	100	–	0	a_4	100	D_1	9
	100	–	1	a_4	100	D_1	10
Y_6	101	–	0	a_5	101	$D_1 D_3$	11
	101	–	1	a_5	101	$D_1 D_3$	12
Y_7	110	–	0	a_5	101	$D_1 D_3$	13
	110	–	1	a_5	101	$D_1 D_3$	14

S_{21}, there is $H_1 = 2$, it means that only 14 rows are in use, whereas there are totally $2^{R_0} = 16$ rows.

For the Mealy FSM $PC_Y Y(\Gamma_9)$ the table of microoperations is represented by Table 5.11. The logic circuit of block BTMS is implemented using embedded memory blocks on the base of Table 5.17

Let us point out that the logic circuit of block BTMS can be implemented using PLAer. In this case the following system of Boolean functions should be constructed:

$$D_r = \bigvee_{h=1}^{H_2} C_{rk} Z_h V_h \quad (r = 1, \ldots, R). \tag{5.37}$$

If the column contains the symbol "–" in the row h of the table of block BTMS, then $V_h = 1$. It allows minimizing system (5.37). For example, $D_1 = F_4 \vee F_9 \vee F_{10} \vee F_{11} \vee F_{12} \vee F_{15} \vee F_{14} = \bar{z}_1 \bar{z}_2 z_3 v_1 \vee z_1 \bar{z}_2 \bar{z}_3 \vee z_1 \bar{z}_2 z_3 \vee \bar{z}_1 z_2 z_3$ (Table 5.18).

Let us discuss an example of logic synthesis for the $PC_Y D(\Gamma_9)$ Mealy FSM having the structural diagram shown in Fig. 5.12. The codes for its collections of microoperations are shown in Table 5.13. Using these codes of collections as well as the state codes from Table 5.15 it is possible to construct the transformed structure of Mealy FSM $PC_Y D(\Gamma_9)$ (Table 5.19). It is constructed in the same way, as it is done for $PC_A Y$ Mealy FSM.

For PD Mealy FSM, the number of bits used in the code $K(Y_t)$ is much more than for equivalent PY Mealy FSM. It means that the logic circuit of block BTSM for PD Mealy FSM should be implemented using LUTs. For the Mealy FSM $PC_Y D(\Gamma_9)$ the table of block BTSM includes $H_0 = 8$ rows (Table 5.20. To implement the logic

Table 5.18   State codes for Mealy FSM $PC_Y D(\Gamma_9)$

a_m	$C(a_s)^t$	α_{t_m}	h
a_1	00000*	$\alpha_{1,1}$	1
a_2	001010	$\alpha_{2,2}$	2
a_3	01000*	$\alpha_{3,3}$	3
a_3	01100*	$\alpha_{4,3}$	4
a_4	10001*	$\alpha_{5,4}$	5
a_4	001011	$\alpha_{2,4}$	6
a_5	00010*	$\alpha_{6,5}$	7
a_5	01011*	$\alpha_{7,5}$	8

Table 5.19   The transformed structure table of Mealy FSM $PC_Y D(\Gamma_9)$

a_m	$K(a_m)$	X_h	Z_h	V_h	h
a_1	000	x_1	$z_3 z_5$	–	1
		$\bar{x}_1$	z_2	–	2
a_2	010	x_2	$z_3 z_5$	–	3
		$\bar{x}_2 x_3$	$z_2 z_3$	–	4
		$\bar{x}_2 x_3$	$z_3 z_5$	v_1	5
a_3	011	x_4	$z_1 z_5$	–	6
		$\bar{x}_1$	z_4	–	7
a_4	100	1	$z_2 z_4 z_5$	–	8
a_5	101	$x_2 x_3$	$z_3 z_5$	–	9
		$x_2 \bar{x}_3$	z_2	–	10
		$\bar{x}_2 x_4$	$z_2 z_4 z_5$	–	11
		$\bar{x}_2 \bar{x}_4$	–	–	12

Table 5.20   Specification of block BTSM for Mealy FSM $PC_Y D(\Gamma_9)$

Y_t	$K(Y_T)$	I_k	$K(I_k)$	a_s	$K(a_s)$	Φ_h	h
Y_1	00000	–	0	a_1	000	–	1
Y_2	00101	I_1	0	a_2	001	D_3	2
		I_2	1	a_4	011	$D_2 D_3$	3
Y_3	01000	–	0	a_3	010	D_2	4
Y_4	01100	–	0	a_3	010	D_2	5
Y_5	10001	–	0	a_4	011	$D_2 D_3$	6
Y_6	00010	–	0	a_5	100	D_1	7
Y_7	01011	–	0	a_6	100	D_1	8

circuit of $PC_Y D$ Mealy FSM, its transformed ST is used to derive systems Z and V, whereas its table for block BTSM is the base to derive system Φ. For example, the following Boolean equation can be derived $D_1 = F_7 \vee F_8 = \bar{z}_1 \bar{z}_2 \bar{z}_3 z_4 \bar{z}_5 \vee \bar{z}_1 z_2 \bar{z}_3 z_4 \bar{z}_5$ from Table 5.20.

References

1. Asahar, P., Devidas, S., Newton, A.: Sequential Logic Synthesis. Kluwer Academic Publishers, Boston (1992)
2. Barkalov, A., Barkalov, A.: Design of mealy finite-state machines with transformation of object codes. Int. J. Appl. Math. Comput. Sci. **15**(1), 151–158 (2005)
3. Barkalov, A., Das, D.: Optimization of logic circuit of microprogrammed mealy automation on PLA. Autom. Control Comput. Sci. **25**(3), 90–94 (1991)
4. Barkalov, A., Titarenko, L.: Logic Synthesis for FSM-Based Control Units. Lecture Notes in Electrical Engineering, vol. 53. Springer-Verlag, Berlin (2009)
5. Barkalov, A., Węgrzyn, M.: Design of Control Units with Programmable Logic. University of Zielona Góra Press (2006)

Chapter 6
Hardware Reduction for Moore EFSMs

Abstract The Chapter is devoted to hardware reduction targeting the elementary LCS-based Moore FSMs. Firstly, the optimization methods are proposed for the base model of EFSM. They are based on the executing either optimal state assignment or transformation of state codes. Two different models are proposed for the case of code transformation. They depend on the numbers of microoperations of FSM and outputs of EMB in use. The models are discussed based on the principle of code sharing. In this case, the state code is represented as a concatenation of the chain code and the code of component inside this chain. The last part of the chapter is devoted to design methods targeting the hybrid FPGAs.

6.1 Optimization of EFSM with the Base Structure

Let us name the Moore FSM PY_E (Fig. 3.13) as an EFSM with the base structure. The structural diagram of FPGA-based PY_E Moore FSM is shown in Fig. 6.1.

The block LUTer (Fig. 6.1) represents the block BIMF (Fig. 3.13). It implements the system of input memory functions (2.9). The block EMBer (Fig. 6.1) represents the block BMO (Fig. 3.13). It implements the system of microoperations (2.16), as well as the function (3.29). We have discussed an example of synthesis for Moore FSM $PY_E(\Gamma_4)$.

To diminish the number of LUTs in the circuit of LUTer, it is necessary to diminish numbers of literals and terms in functions (2.9). Two methods can be used for optimizing LUTer of PY_E FSM:

1. Optimal state assignment.
2. Transformation of state codes into class codes.

Both these methods are based on existence of classes of PES.

Let us discuss these methods for the case of GSA Γ_{10} (Fig. 6.2). It is marked by states of Moore FSM using the rules [2]. The following sets and their parameters can be derived from GSA Γ_{10}: $X = \{x_1, \ldots, x_4\}$, $L = 4$, $Y = \{y_1, y_5\}$, $N = 5$, $A = \{a_1, \ldots, a_{19}\}$, $M = 19$, $R = 5$, $T = \{T_1, \ldots, T_5\}$, $\Phi = \{D_1, \ldots, D_5\}$.

© Springer International Publishing AG 2018

A. Barkalov et al., *Logic Synthesis for Finite State Machines Based on Linear Chains of States*, Studies in Systems, Decision and Control 113, DOI 10.1007/978-3-319-59837-6_6

Fig. 6.1 Structural diagram
of FPGA-based PY_E Moore
FSM

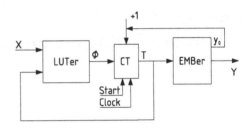

The optimal state assignment leads to P_0Y_E Moore FSM. It has the same structure as PY_E FSM (Fig. 6.1). The following method can be used for synthesis of P_0Y_E Moore FSM:

1. Marking the initial GSA Γ and creating the set of states A.
2. Constructing the set of ELCS C_E.
3. Constructing the partition $\Pi_{CE} = \{B_1, \ldots, B_{IE}\}$.
4. Executing the optimal state assignment.
5. Constructing the table of LUTer.
6. Constructing the table of EMBer.
7. Implementing the FSM logic circuit.

Let us apply the procedure P_1 to the GSA Γ_{10}. It produces the set $C_E = \{\alpha_1, \ldots, \alpha_7\}$ with the following elementary LCSs: $\alpha_1 = \langle a_1, a_2, a_3 \rangle$, $\alpha_2 = \langle a_4, \ldots, a_7 \rangle$, $\alpha_3 = \langle a_5, a_9, a_{10} \rangle$, $\alpha_4 = \langle a_{11} \rangle$, $\alpha_5 = \langle a_{12}, a_{13}, a_{14} \rangle$, $\alpha_6 = \langle a_{15}, \ldots, a_{18} \rangle$, and $\alpha_7 = \langle a_{15} \rangle$. There is $G1 = 7$ in the discussed case.

To construct the partition Π_{CE}, let us form the set $O_E(\Gamma_{10})$. This set includes outputs of ELCS $\alpha_g \in C_E$. Using the Definition 3.4, the following set can be found: $O_E(\Gamma_{10}) = \{a_3, a_7, a_{10}, a_{11}, a_{14}, a_{18}, a_{19}\}$. The following classes of PES can be found for the states $a_m \in O_E(\Gamma_{10})$: $\{a_3\}$, $\{a_7, a_{10}\}$, $\{a_{14}, a_{18}\}$, and a_{19}. It gives the following partition $\Pi_{CE} = \{B_1, \ldots, B_5\}$ with the classes $B_1 = \{\alpha_1\}$, $B_1 = \{\alpha_2, \alpha_3\}$, $B_3 = \{\alpha_4\}$, $B_4 = \{\alpha_5, \alpha_6\}$, and $B_5 = \{\alpha_7\}$.

The aim of optimal state assignment is to find such state codes that the class codes $K(B_i)$ will be represented by the minimum possible amount of cubes of R-dimensional Boolean space. Let us point out that the initial state $a_i \in A$ should have code with all zeroes. One of the possible variants is shown in Fig. 6.3.

This state assignment is a natural state assignment satisfying to (3.26). It can be found from Fig. 6.3 that: the class B_1 is determined by the cube 000**, the class B_2 by 0*1**, the class B_3 by 1*1**, the class B_4 by *10**, and the class B_5 by the 100**. It gives the following class codes: $K(B_1) = 000**$, $K(B_2) = 0*1**$, $K(B_3) = 100**$, $K(B_4) = *10**$ and $K(B_5) = 1*1**$.

The table of LUTer includes the following columns: B_i, $K(B_i)$, a_s, $K(a_s)$, X_h, Φ_h, h. To construct this table a system of generalized formulae of transitions (GFT) [4, 7] should be formed. In the discussed case, it is the following system:

Fig. 6.2 Initial
graph-scheme of algorithm
Γ_{10}

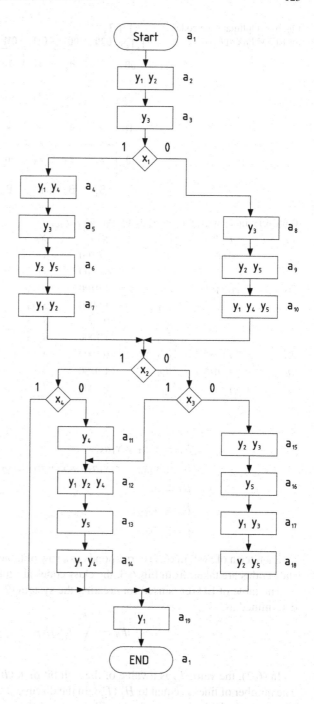

Fig. 6.3 Optimal state codes for EFSM $P_0 Y_E(\Gamma_{10})$

$T_4 T_5$ \ $T_1 T_2 T_3$	000	001	011	011	110	111	101	100
00	a_1	a_4	a_8	a_{12}	a_{15}	*	a_{19}	a_{11}
01	a_2	a_5	a_9	a_{13}	a_{16}	*	*	*
11	*	a_7	*	*	a_{18}	*	*	*
10	a_3	a_6	a_{10}	a_{14}	a_{17}	*	*	*
	B_1	B_2	B_2	B_4	B_4		B_5	B_3

Table 6.1 Table of LUTer for Mealy EFSM $P_0 Y_E(\Gamma_{10})$

B_i	$K(B_i)$	a_s	$K(a_s)$	X_h	Φ_h	h
B_1	0 00**	a_4	0 0100	x_1	D_3	1
		a_8	0 1100	$\bar{x}_1$	$D_2 D_3$	2
B_2	0 *1**	a_{19}	1 0100	$x_2 x_4$	$D_1 D_3$	3
		a_{11}	1 0000	$x_2 \bar{x}_4$	D_1	4
		a_{12}	0 1000	$\bar{x}_2 x_3$	D_2	5
		a_{15}	1 1000	$\bar{x}_2 \bar{x}_3$	$D_1 D_2$	6
B_3	1 00**	a_{12}	0 1000	1	D_2	7
B_4	* 10**	a_{19}	1 0100	1	$D_1 D_3$	8
B_5	1 *1**	a_1	0 0000	1	–	9

$$B_1 = x_1 a_4 \vee \bar{x}_1 a_8;$$
$$B_2 = x_2 x_4 \vee x_2 \bar{x}_4 a_{11} \vee \bar{x}_2 x_3 a_{12} \vee \bar{x}_2 \bar{x}_3 a_{15};$$
$$B_3 = a_{12}; \tag{6.1}$$
$$B_4 = a_{19};$$
$$B_5 = a_1.$$

Each term of system (6.1) corresponds to a row of table of LUTer (Table 6.1). The state codes are taken from Fig. 6.3, the class codes are found before.

The table of LUTer is used for creating the system (2.9). The term F_h of (2.9) is determined as:

$$F_h = \left(\bigvee_{r=1}^{R} T_r^{l_{ir}} \right) X_h. \tag{6.2}$$

In (6.2), the value l_{ir} is a value of the r-th bit of $K(B_i)$ from the line number h. The number of lines is equal to $H_E^0(\Gamma_j)$. In the discussed case, there is $H_E^0(\Gamma_{10}) = 9$. The variable l_{ir} belongs to the set $\{0, 1, *\}$ and there are $T_r^0 = \bar{T}_r$, $T_r^1 = T_r$ and $T_r^* = 1 (r = 1, \ldots, R)$. The functions (2.9) are determined as

$$F_r = \bigvee_{h=1}^{H_E^0} C_{rh} F_h \quad (r = 1, \ldots, R).$$ (6.3)

In (6.3), the Boolean variable $C_{rh} = 1$ only if the function D_r is written in the line h of the table of LUTer.

For example, the function $D_1 \in \Phi$ is written in the lines 3, 4, 6 and 8 of Table 6.1. After minimizing, the following function can be found:

$$D_1 = \bar{T}_1 T_3 x_3 \vee \bar{T}_1 T_3 \bar{x}_3 \vee T_2 \bar{T}_3.$$ (6.4)

Using the same approach, the functions D_2 and D_3 can be found:

$$D_2 = \bar{T}_1 \bar{T}_2 \bar{T}_3 \bar{x}_1 \vee \bar{T}_1 T_3 \bar{x}_2 \vee T_1 \bar{T}_2 \bar{T}_3,$$ (6.5)

$$D_3 = \bar{T}_1 \bar{T}_2 \bar{T}_3 \vee \bar{T}_1 T_3 x_2 x_4 \vee T_2 \bar{T}_3.$$ (6.6)

Let us point out that each of functions (6.4)–(6.6) can be implemented using only a single LUT having $S = 5$. In the discussed case, the system (2.9) depends only on state variables T_1–T_3. So, it is the function:

$$\Phi = \Phi(T^1, X),$$ (6.7)

where $T^1 \subseteq T$. At the same time, only the input memory functions D_1–D_3 should be formed by the LUTer. It is the best possible solution.

In the common case, the best solution is determined by the equation

$$|T^1| = \lceil \log_2 |\Pi_{CE}| \rceil.$$ (6.8)

The table of EMBer is constructed in a trivial way. In includes the columns $K(a_m)$, $Y(a_m)$, h. In the common case, this table includes H_{EMB} rows, where

$$H_{EMB} = 2^R.$$ (6.9)

Let us point out that only M rows include the collections of microoperations.

If a state $a_m \in A$ is not the output of ELCs, then the corresponding cell should include the variable y_0. This variable should be included in the corresponding line of the table of EMBer. In the discussed case, there is $H_{EMB} = 32$. The first 8 rows of the table of EMBer is represented by Table 6.2 for Moore EFSM $P_0 Y_E(\Gamma_{10})$.

The column m is added to show the correspondence among the rows of table of EMBer and the states $a_m \in A$. Let the following condition take place:

$$2^R(N + 1) \leq V_0.$$ (6.10)

Table 6.2 The part of table of EMBer for Mealy UFSM $P_0 Y_E(\Gamma_{10})$

$K(a_m) T_1 \ldots T_5$	$Y(a_m) y_0 \ldots y_5$	h	m
0 0000	10 0000	1	1
0 0001	11 1000	2	2
0 0010	00 0100	3	3
0 0011	00 0000	4	*
0 0100	11 0010	5	4
0 0101	10 0100	6	5
0 0110	10 1001	7	6
0 0111	01 1000	8	7

The symbol V_0 stands for the number of cells of EMB under $t_F = 1$. If the condition (6.10) takes place, the only one EMB is enough for implementing the logic circuit of EMBer.

In the discussed case, the left part of (6.10) produces the value $186 = 32 * 6$. Let it be the EMB with configuration 32×8 in a particular FPGA chip. Let this chip include LUTs having five inputs. In this case, it is enough a single EMB for implementing the circuit of EMBer. Each function (6.4)–(6.6) is implemented using a single LUT. To implement the circuit of counter CT, five LUTs are necessary. So, the logic, circuit of Moore EFSM $P_0 Y_E(\Gamma_{10})$ includes 8 LUTs and a single EMB (Fig. 6.4). The counter is shown as a single block. We do not discuss the organization of counters in our book. It can be found in many books connected with logic design, for example in [10, 11].

Let us point out that there are GSAs for which the condition (6.8) does not take place. In this case, the hardware reduction can be executed due to transformation of state codes into class codes. It leads to $P_C Y$ Moore FSM (Fig. 3.10). Let us discuss this approach for ELCS-based Moore FSMs.

Let us find the partition $\Pi_{CE} = \{B_1, \ldots, B_{IE}\}$ for a given GSA Γ_j. Let us encode the classes $B_i \in \Pi_{CE}$ by binary codes $K(B_i)$ having R_{CE} bits:

$$R_{CE} = \lceil \log_2 I E \rceil. \tag{6.11}$$

Let us use the variables $\tau_r \in \tau$ for encoding the classes.

Let the following condition take place:

$$2^R (N + R_{CE} + 1) \leq V_0. \tag{6.12}$$

In this case, the functions (3.17) are generated by the EMBer. It results in $P_{C1} Y_E$ Moore FSM (Fig. 6.5).

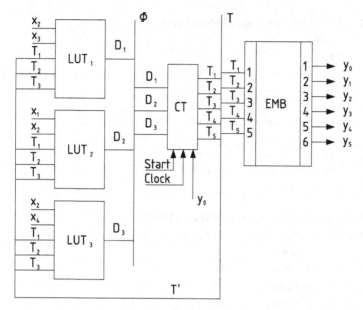

Fig. 6.4 Logic circuit of Moore EFSM $P_0Y_E(\Gamma_{10})$

Fig. 6.5 Structural diagram of FPGA=based $P_{C1}Y_E$ Moore FSM

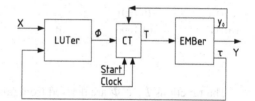

In this model, the LUTer generates functions (3.15), whereas the EMBer functions (2.16), (3.15) and (3.29). The method of synthesis for $P_{C1}Y_E$ Moore FSM is similar to the method for P_0Y_E FSM. There is the following difference:

4. Executing the natural state assignment.
4a. Encoding of the classes $B_i \in \Pi_{CE}$.

Let us discuss an example of synthesis for Moore EFSM $P_{C1}Y_E(\Gamma_{10})$. The sets C_E and Π_{CE} are already found. Let us use the same state codes as the ones from Fig. 6.3.

There is $IE = 5$ in the discussed case. It gives the value $R_{CE} = 3$. But the analysis of Table 6.1 shows that the transitions are executed automatically for the states $a_m \in B_5$. It means that only classes $B_1, \ldots, B_4$ should be encoded. It gives the set $\tau = \{\tau_1, \tau_2\}$. Let us encode the classes in the trivial way: $K(B_1) = 00, \ldots, K(B_4) = 11$. The table of LUTer is constructed on the base of the system of GFT. In the discussed case, it is Table 6.3 constructed using the system (6.1).

Table 6.3 Table of LUTer for Moore EFSM $P_{C1}Y_E(\Gamma_{10})$

B_i	$K(B_i)$	a_s	$K(a_s)$	X_h	Φ_h	h
B_1	00	a_4	0 0100	x_1	D_3	1
		a_8	0 1100	$\bar{x}_1$	$D_2 D_3$	2
B_2	01	a_{19}	1 0100	$x_2 x_4$	$D_1 D_3$	3
		a_{11}	1 0000	$x_2 \bar{x}_4$	D_2	4
		a_{12}	0 1000	$\bar{x}_2 \bar{x}_3$	D_2	5
		a_{15}	1 1000	$\bar{x}_2 \bar{x}_3$	$D_1 D_2$	6
B_3	10	a_{12}	0 1000	1	D_2	7
B_4	11	a_{19}	1 0100	1	$D_1 D_3$	8

Table 6.4 The part of table of EMBer for Moore EFSM $P_{C1}Y_E(\Gamma_{10})$

$K(a_m) T_1 \ldots T_5$	$Y(a_m) y_0 \ldots y_5$	$K(B_i) \tau_1 \tau_2$	h	m
0 0000	10 0000	00	1	1
0 0001	11 1000	00	2	2
0 0010	00 0100	00	3	3
0 0011	00 0000	00	4	*
0 0100	11 0010	00	5	4
0 0101	10 0100	00	6	5
0 0110	10 1001	00	7	6
0 0111	01 1000	01	8	7

The functions $D_r \in \Phi$ are derived from this table. They depend on the terms

$$F_h = (\bigwedge_{r=1}^{R_{CE}} \tau_r^{l_{ir}})x_h \quad (h = 1, \ldots, H_E^C). \tag{6.13}$$

The meaning is obvious for each element of (6.13).

After minimizing, the following functions can be derived from Table 6.3

$$\begin{aligned}
D_1 &= \bar{\tau}_1 \tau_2 x_2 \vee \bar{\tau}_1 \tau_2 \bar{x}_3 \vee \tau_1 \tau_2; \\
D_2 &= \bar{\tau}_1 \bar{\tau}_2 \bar{x}_1 \vee \bar{\tau}_1 \tau_2 \bar{x}_2 \vee \tau_1 \bar{\tau}_2; \\
D_3 &= \bar{\tau}_1 \bar{\tau}_2 \vee \bar{\tau}_1 \tau_2 x_2 x_4 \vee \tau_1 \tau_2.
\end{aligned} \tag{6.14}$$

Analysis of the system (6.14) shows that each its equations can be implemented using LUTs with $S = 4$.

The table of EMBer includes the column $K(B_i)$. If $a_m = O_g$, then the cell with address $K(a_m)$ contains the code $K(B_i)$ such that $\alpha_g \in B_i$. The first 8 rows for this table are shown in Table 6.4. The state a_8 is the output of ELCS $\alpha_2 \in B_2$. Because of it, the code 01 is placed in the row 8 of Table 6.4.

Fig. 6.6 Structural diagram
of FPGA-based $P_{C2}Y_E$
Moore FSM

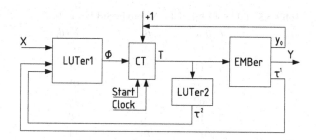

If the condition (6.12) is violated, then a part of the code $K(B_i)$ should be generated by the block of transformer of state codes BTC (Fig. 3.10). It leads to $P_{C2}Y_E$ Moore FSM (Fig. 6.6).

In this model, the LUTer1 generates the functions (3.15), the EMBer generates functions (2.16), (2.28) and

$$\tau^1 = \tau^1(T). \tag{6.15}$$

The block LUTer2 implements functions

$$\tau^2 = \tau^2(T). \tag{6.16}$$

Obviously, the following condition takes place:

$$\tau^1 \cup \tau^2 = \tau; \\ \tau^1 \cap \tau^2 = \emptyset; \tag{6.17}$$

The following method can be used for synthesis of $P_{C2}Y_E$ Moore FSM:

1. Creating the set of states A.
2. Constructing the set of ELCS C_E.
3. Constructing the set $\Pi_{CE} = \{B_1, \ldots, B_{IE}\}$.
4. Executing the natural state assignment.
5. Encoding of the classes $B_i \in \Pi_{CE}$.
6. Constructing the table of LUTer1.
7. Constructing the table of LUTer2.
8. Constructing the table of EMBer.
9. Implementing the FSM logic circuit.

Let us discuss an example of synthesis for Moore EFSM $P_{C2}Y_E(\Gamma_{10})$. The sets C_E, Π_{CE}, τ are already found, as well as the codes of states $a_m \in A$ and classes $B_i \in \Pi_{CE}$. The table of LUTer1 is the same as Table 6.3.

Table 6.5 Table of LUTer2 for Moore EFSM $P_{C2}Y_E10$

a_m	$K(a_m)$	B_i	$K(B_i)$	τ_m^2	m
a_3	0 0010	B_1	00	–	3
a_7	0 0111	B_2	01	τ_2	7
a_{10}	0 1110	B_2	01	τ_2	10
a_{11}	1 0000	B_3	10	–	11
a_{14}	0 1010	B_4	10	–	14
a_{18}	1 1011	B_4	11	τ_2	18

The number of cells is equal 2^R for EMBs of EMBer. It allows finding the number of outputs t_F. Let the following conditions take places:

$$t_F > N + 1; \tag{6.18}$$

$$t_F < N + R_{CE} + 1. \tag{6.19}$$

In this case, the set τ should be derived using (6.17). The set τ^1 includes R_{CE1} elements, where

$$R_{CE1} = t_F - (N + 1). \tag{6.20}$$

The set τ^2 includes R_{CE2} elements, where

$$R_{CE2} = R_{CE} - R_{CE1}. \tag{6.21}$$

Let the following condition take place:

$$S \geq R. \tag{6.22}$$

In (6.22), the symbol S stands for the number of inputs of LUTs used for design of an FSM circuit. In this case, the set τ can be divided in an arbitrary way.

Let the configuration 32×7 exist for an EMB in use. So, there is $t_f = 7$. Because of $N = 5$, $R_{CE} = 2$, the conditions (6.18)–(6.19) take places. Using (6.20)–(6.21), the following equality $R_{CE1} = R_{CE2} = 1$ can be found. Let it be $S = 5$. So, the condition (6.22) takes place. It means that the following sets can be formed $\tau^1 = \{\tau_1\}$ and $\tau^2 = \{\tau_2\}$.

The table of LUTer2 is constructed only for outputs of ELCS $\alpha_g \in C_E$. It contains the columns a_m, $K(a_m)$, B_i, $K(B_i)$, τ_m^2, m. The class $B_i \in \Pi_{CE}$ is placed in the row number m if $(a_m = O_g)\&(\alpha_g \in B_i) = 1$. The column τ_m^2 includes the variables $\tau_r \in \tau^2$ corresponding to ones in the code $K(B_i)$ from the row number m $(m = 1, \ldots, M)$. The table of LUTer2 includes 6 rows in the discussed case (Table 6.5).

This table is used to program a LUT implementing the function $\tau_2 = \tau_2(T)$. If the condition (6.22) is violated, then the equations for all functions $\tau_r \in \tau$ should be obtained. The set τ^2 should include the functions $\tau_r \in \tau$ corresponding to logic

T_1T_2＼$T_1T_2T_3$	000	001	011	010	110	111	101	100
00	*	*	*	*	*	*	*	0
01	*	*	*	*	*	*	*	*
11	*	1	*	*	1	*	*	*
10	0	*	1	0	*	*	*	*

Fig. 6.7 Karnaugh map for function τ_2

circuit with minimum number of LUTs. We do not discuss this problem in our book. Let us only point out that codes of states $a_m \notin O_E$ can be considered as "don't care" for functions $\tau_r \in \tau$. This property can be used for minimizing these functions. To illustrate this issue, let us form a Karnaugh map for function τ_2 (Fig. 6.7). It includes the signs * for all states $a_m \in O_E(\Gamma_{10})$. The following equation can be obtained from this map:

$$\tau_2 = T_3 \vee T_5. \tag{6.23}$$

Obviously, the corresponded circuit is implemented using only a single LUT with $S = 2$.

6.2 Synthesis of EFSM with Code Sharing

The structural diagram of PY_{EC} Moore FSM is shown in Fig. 3.18. We discuss FPGA-based structures of FSM circuits. In the case of PY_{EC} Moore FSM, the block BIMF is represented by LUTer and the block BMO is represented by EMBer (Fig. 6.8).

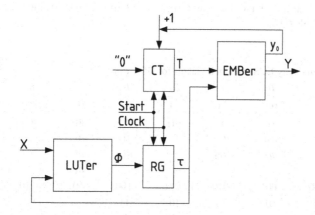

Fig. 6.8 Structural diagram of FPGA-based PY_{EC} Moore FSM

In this FSM, the LUTer forms functions

$$\Psi = \Psi(\tau, X). \qquad (6.24)$$

These functions are used for loading codes $K(\alpha_g)$ of ELCS $\alpha_g \in C_E$ into the register RG. The state variables $T_r \in T$ are used for encoding of the states $a_m \in A(\alpha_g)$, where $r = 1, \ldots, R_{G1}$. The chain variables $\tau_r \in \tau$ are used for encoding of the ELCS $\alpha_g \in C_E$, where $r = 1, \ldots, R_{G1}$. The value of R_{G1} is determined by (3.64). The states are encoded by codes $C(a_m)$, the chains are encoded by codes $K(\alpha_g)$. The code $K(a_m)$ is determined as the concatenation (3.66). This code is considered as an address of a memory cell.

The proposed design method for PY_{EC} Moore FSM includes the following steps:

1. Creating the set of states A.
2. Constructing the set of ELCS C_E.
3. Encoding of ELCS $\alpha_g \in C_E$.
4. Encoding of states $a_m \in A(\alpha_g)$ $(g = 1, \ldots, G1)$.
5. Constructing the table of LUTer.
6. Constructing the table of EMBer.
7. Implementing the FSM logic circuit.

Let us discuss an example of synthesis for Moore EFSM $PY_{EC}(\Gamma_{10})$. The steps 1 and 2 are already executed. Let us construct Table 6.6 with ELCS $\alpha_g \in C_E$ and classes $B_i \in \Pi_{CE}$. This table can be constructed using the previous results. Table 6.6 also includes state codes $C(a_m)$, chain codes $K(\alpha_g)$ and class codes $K(B_i)$. We will discuss these codes a bit later.

It is found before that there are $R_{G1} = 3$ and $R_{C1} = 2$. So, there are the sets $\tau = \{\tau_1, \tau_2, \tau_3\}$ and $T = \{T_1, T_2\}$. Let us execute the steps 3 and 4 of the proposed method.

Let us encode the ELCS $\alpha_g \in C_E$ in the trivial way: $K(\alpha_1) = 000$, $K(\alpha_2) = 001, \ldots, K(\alpha_7) = 110$ (Table 6.6). To satisfy (3.65), the first components of all ELCS $\alpha_g \in C$ should have the code $C(a_m) = 00$, the second 01, the third 10 and the fourth 11 (Table 6.6).

Table 6.6 Elementary LCSs and their classes for GSA Γ_{10}

B_i	B_1	B_2		B_3	B_4		B_5	$C(a_m)$
α_g	α_1	α_2	α_3	α_4	α_5	α_6	α_7	
a_m	a_1	a_4	a_8	a_{11}	a_{12}	a_{15}	a_{19}	00
	a_2	a_5	a_9	–	a_{13}	a_{16}	–	01
	a_3	a_6	a_{10}	–	a_{14}	a_{17}	–	10
	–	a_7	–	–	–	a_{18}	–	11
$K(\alpha_g)$	000	001	010	011	100	101	110	–
$K(B_i)$	00	01		10	11		*	–

This table gives the codes (3.66). For example the following codes can be found: $K(a_1) = 00000$, $K(a_2) = 01001$, $K(a_3) = 01010$, and so on.

To construct the table of LUTer, the set of outputs O_E should be found. In the discussed case, the set $O_E(\Gamma_{10}) = \{a_3, a_7, a_{10}, a_{11}, a_{14}, a_{18}, a_{19}\}$. Let us form the system of formulae of transitions for the states $a_m \in O_E(\Gamma_{10})$. It is the following system:

$$a_3 \rightarrow x_1 a_4 \vee \bar{x}_1 a_8;$$
$$a_7 \rightarrow x_2 x_4 a_{19} \vee x_2 \bar{x}_4 a_{11} \vee \bar{x}_2 x_3 a_{12} \vee \bar{x}_2 \bar{x}_3 a_{15};$$
$$a_{10} \rightarrow x_2 x_4 a_{19} \vee x_2 \bar{x}_4 a_{11} \vee x_2 x_3 a_{12} \vee \bar{x}_2 \bar{x}_3 a_{15};$$
$$a_{11} \rightarrow a_{12}; \qquad\qquad\qquad\qquad\qquad\qquad\qquad (6.25)$$
$$a_{14} \rightarrow a_{19};$$
$$a_{18} \rightarrow a_{19};$$
$$a_{19} \rightarrow a_1.$$

The table of LUTer includes the following columns: α_m, $K(\alpha_m)$, α_s, $K(\alpha_s)$, X_h, Ψ_h, h. Each line of the table corresponds to one term of SFT. Each state $a_i \in A$ is replaced by an ELCS $\alpha_g \in C_E$ such that $a_i \in A(\alpha_g)$. The table contains $H_{EC}(\Gamma_j)$ lines. In the case of $PY_{EC}(\Gamma_{10})$, the table of LUTer includes $H_{EC}(\Gamma_{10}) = 14$ lines (Table 6.7).

Table 6.7 Table of LUTer of Moore EFSM $PY_{CE}(\Gamma_{10})$

α_m	$K(\alpha_m)$	α_s	$K(\alpha_s)$	X_h	Ψ_h	h
α_1	000	α_2	001	x_1	D_3	1
		α_3	010	$\bar{x}_1$	D_2	2
α_2	001	α_7	110	$x_2 x_4$	$D_1 D_2$	3
		α_4	011	$x_2 \bar{x}_4$	$D_2 D_3$	4
		α_5	100	$\bar{x}_2 x_3$	D_1	5
		α_6	101	$\bar{x}_2 \bar{x}_3$	$D_1 D_3$	6
α_3	010	α_7	110	$x_2 x_4$	$D_1 D_2$	7
		α_4	011	$x_2 \bar{x}_4$	$D_2 D_3$	8
		α_5	100	$\bar{x}_2 x_3$	D_1	9
		α_6	101	$\bar{x}_2 \bar{x}_3$	$D_1 D_3$	10
α_4	011	α_5	100	1	D_1	11
α_5	100	α_7	110	1	$D_1 D_2$	12
α_6	101	α_7	110	1	$D_1 D_2$	13
α_7	110	α_1	000	1	–	14

The connection is obvious between the system (6.25) and Table 6.7. The chain codes are taken from Table 6.6. This table is the base for constructing the system (6.24). Terms of (6.24) are determined by the following expression:

$$F_h = (\bigwedge_{r=1}^{R_{G1}} \tau_r^{l_{gr}}) X_h \quad (h = 1, \dots, H_{EC}). \tag{6.26}$$

The symbol l_{gr} stands for the value of r-th bit of the code $K(\alpha_g)$ from the h-th row of the table.

For the example, the following equation can be derived from Table 6.7:

$$D_1 = \bar{\tau}_1 \bar{\tau}_2 \tau_3 (x_2 \vee \bar{x}_3) \vee \bar{\tau}_1 \tau_2 \bar{\tau}_3 (x_2 \vee \bar{x}_3) \vee \tau_2 \tau_3 \vee \tau_1 \bar{\tau}_2 \bar{\tau}_3 \vee \tau_1 \tau_3. \tag{6.27}$$

To get this equation, the code 111 was used for minimizing the function D_1. To implement the logic circuit corresponding to (6.27), it is enough one LUT having $S = 5$.

The table of EMBer is constructed in the same way as for PY_E Moore EFSM. The codes $K(a_m)$ are the same for both $PY_E(\Gamma_{10})$ and $PY_{CE}(\Gamma_{10})$.

To optimize the circuit of LUTer, two methods can be used:

1. The optimal chain assignment ($P_0 Y_{CE}$ EFSM).
2. The transformation of chain codes ($P_C Y_{CE}$ EFSM).

Both approaches are based on constructing the partition Π_{CE}. Let us discuss this methods.

The following approach can be used for synthesis of $P_0 Y_{EC}$ Moore EFSM:

1. Creating the set of states A.
2. Constructing the set of ELCS C_E.
3. Encoding of states $a_m \in A(\alpha_g)$.
4. Constructing the partition Π_{CE}.
5. Optimal chain assignment.
6. Constructing the table of LUTer.
7. Constructing the table of EMBer.
8. Implementing the FSM logic circuit.

Obviously, the structural diagrams are the same for PY_{EC} and $P_0 Y_{EC}(\Gamma)$ FSM. The sets A, Π_{CE}, $B_i \in \Pi_{CE}$ are already found, as well as the state code $C(a_m)$. They are represented by Table 6.6.

The optimal chain assignment should result in such chain codes that each class $B_i \in \Pi_{CE}$ is represented by minimum possible amount of cubes of R_{G1}-dimensional Boolean space. One of the possible variants is shown in Fig. 6.9. The following codes can be derived from Fig. 6.9: $K(B_1) = 00*$, $K(B_2) = 01*$, $K(B_3) = 110$, $K(B_4) = 10*$ and $K(B_5) = 111$. Using chain codes from Fig. 6.9 and state codes from Table 6.6 the state codes (3.66) can be found. They are shown in Fig. 6.10.

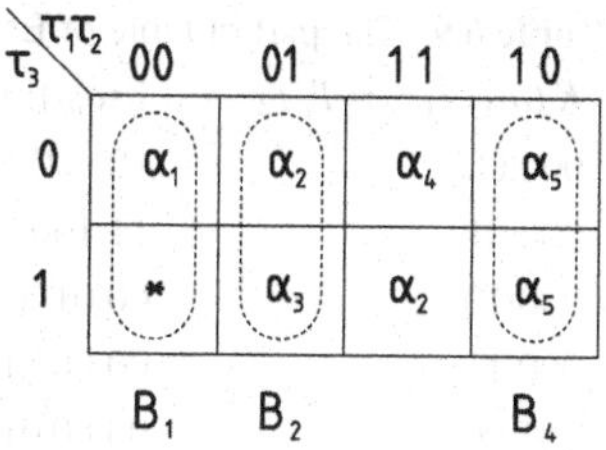

Fig. 6.9 Optimal chain codes for Moore EFSM $P_0 Y_{EC}(\Gamma_{10})$

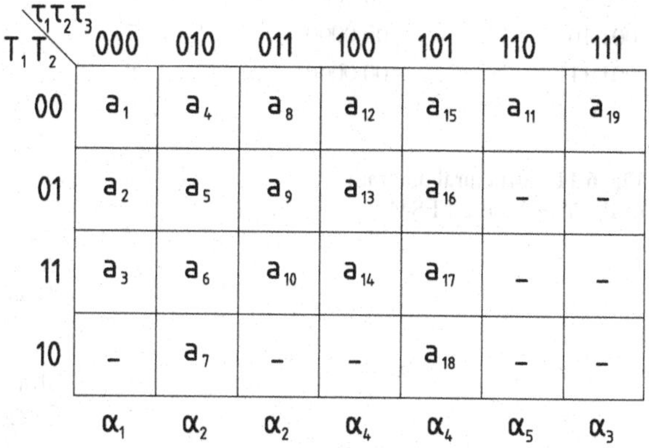

Fig. 6.10 State codes for Moore EFSM $P_0 Y_{EC}(\Gamma_{10})$

Table 6.8 Table of LUTer of Moore EFSM $P_0 Y_{CE}(\Gamma_{10})$

B_i	$K(B_i)$	a_s	$K(a_s)$	X_h	Φ_h	h
B_1	00*	a_4	0 1000	x_1	D_2	1
		a_8	0 1100	$\bar{x}_1$	$D_2 D_3$	2
B_2	01*	a_{19}	1 1100	$x_2 x_4$	$D_1 D_2 D_3$	3
		a_{11}	1 1100	$x_2 \bar{x}_4$	$D_1 D_2$	4
		a_{12}	1 0000	$\bar{x}_2 x_3$	D_1	5
		a_{15}	1 0100	$\bar{x}_2 \bar{x}_3$	$D_1 D_3$	6
B_3	110	a_{12}	1 0000	1	D_1	7
B_4	11*	a_{19}	1 1100	1	$D_1 D_2 D_3$	8

The table of LUTer includes the following columns: B_i, $K(B_i)$, a_s, $K(a_s)$, X_h, Φ_h, h. It is constructed using the system of generalized formulae of transitions. In the discussed case, the system of GFT is represented by (6.1). The table of LUTer is represented by Table 6.8.

The table of LUTer is used for constructing the system (6.24). The terms of (6.24) are represented by (6.28):

$$F_h = (\bigwedge_{r=1}^{R} \tau_r^{l_{ir}}) G1_h \quad (h = 1, \ldots, H_{EC}^0). \tag{6.28}$$

Table 6.9 The part of table of EMBer for Moore EFSM $P_0 Y_{EC}(\Gamma_{10})$

$K(a_m)\,\tau_1\tau_2\tau_3 T_1 T_2$	$Y(a_m)\,y_0\ldots y_5$	h	m
000 00	10 0000	1	1
000 01	11 1000	2	2
000 10	00 0100	3	3
000 11	00 0000	4	*
001 00	00 0010	5	*
001 01	00 0000	6	*
001 10	00 0000	7	*
001 11	00 0000	8	*

Fig. 6.11 Structural diagram
of $P_{C1} Y_{EC}$ Moore EFSM

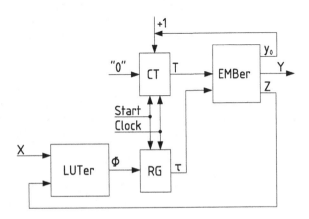

The symbol l_{ir} stands for the value of the r-th bit of the code $K(B_i)$ from the h-th row of the table derived from Table 6.8:

$$D_1 = \bar{\tau}_1\tau_2 \vee \tau_1\tau_2\bar{\tau}_3 \vee \tau_1\bar{\tau}_2;$$
$$D_2 = \bar{\tau}_1\bar{\tau}_2 \vee \bar{\tau}_1\tau_2 x_2 \vee \tau_1\bar{\tau}_2 = \bar{\tau}_2 \vee \bar{\tau}_1\tau_2 x_2; \qquad (6.29)$$
$$D_3 = \bar{\tau}_1\bar{\tau}_2\bar{x}_1 \vee \bar{\tau}_1\tau_2 x_2 x_4 \vee \bar{\tau}_1\tau_2\bar{x}_2\bar{x}_3 \vee \tau_1\bar{\tau}_2.$$

The functions D_1, D_2 can be implemented using one LUT having $S = 3$. It is necessary to have five inputs in a LUT implementing the circuit for D_3 (Fig. 6.11).

The table of EMBer has the same columns as for the case of $P_0 Y_E$. The first 8 lines are shown in Table 6.9 for the discussed case. Let us point out that the codes 001^{**} are not used as state codes $K(a_m)$ (see Fig. 6.9).

Using transformation of codes $K(\alpha_g)$ into codes $K(B_i)$ leads to $P_{C2} Y_{EC}$ Moore EFSM (Fig. 6.12).

The functions Z are used for creating the input memory functions

$$\Phi = \Phi(Z, X). \qquad (6.30)$$

Fig. 6.12 Structural diagram of $P_{C2}Y_{EC}$ Moore EFSM

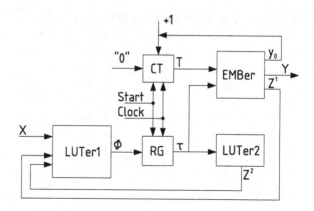

The set Z includes R_{CE} variables, where the value of R_{CE} is determined by (6.11).

The design methods include the same steps for EFSMs $P_{C1}Y_E$ and $P_{C1}Y_{EC}$. Let us discuss an example of synthesis for Moore EFSM $P_{C1}Y_E(\Gamma_{10})$. There are the sets A, Π_{CE} and $B_i \in \Pi_E$ for this example (Table 6.6). Table 6.6 also includes the codes $C(a_m)$ and $K(\alpha_g)$ for the discussed case. It gives the state codes $K(a_m)$ represented as (3.66).

The transitions are not included into the table of LUTer for the class $B_5 \in \Pi_{CE}$. It means that the formula (6.11) can be refined for this case:

$$R_{CE} = \lceil \log_2(IE - 1) \rceil. \tag{6.31}$$

The value $R_{CE} = 2$ is obtained for the discussed case. It gives the set $Z = \{z_1, z_2\}$. Let us encode the class $B_i \in \Pi_{CE}$ using the codes from Table 6.6.

The table of LUTer is constructed using the system of GFT. In the discussed case, it is the system (6.1). The table of LUTer includes the same columns as its counterpart for P_0Y_{CE} EFSM. In the discussed case, it is Table 6.10. This table is used for constructing the functions (6.30). These functions depend on product terms

$$F_h = (\bigwedge_{r=1}^{R_{CE}} z_r^{l_{ir}})X_h \quad (h = 1, \ldots, H_{EC}^{C1}). \tag{6.32}$$

The table of LUTer of Moore EFSM $P_{C1}Y_{CE}(\Gamma_{10})$ includes $H_{CE}^{C1}(\Gamma_{10}) = 8$ rows (Table 6.10). The functions (6.33) are derived from Table 6.10.

$$D_1 = \bar{z}_1 z_2 \bar{x}_1 \vee \bar{z}_1 z_2 \bar{x}_2 \vee z_1;$$
$$D_2 = \bar{z}_1 \bar{z}_2 \bar{x}_1 \vee \bar{z}_1 z_2 x_2 \vee z_1 z_2; \tag{6.33}$$
$$D_3 = \bar{z}_1 \bar{z}_2 x_1 \vee \bar{z}_1 z_2 x_2 \bar{x}_2 \vee \bar{z}_1 z_2 \bar{x}_2 \bar{x}_3.$$

The table of EMBer is constructed as its counterpart for $P_C Y_1 E$ Moore EFSM. The part of this table (Table 6.11) is shown for the discussed example.

Table 6.10 Table of LUTer of Moore EFSM $P_{C1}Y_{CE}(\Gamma_{10})$

B_i	$K(B_i)$	a_s	X_h	Φ_h	h
B_1	00	a_4	0 0100	x_1	D_3	1
		a_8	0 1000	$\bar{x}_1$	D_2	2
B_2	01	a_{19}	1 1000	$x_2 x_4$	$D_1 D_2$	3
		a_{11}	0 1100	$x_2 \bar{x}_4$	$D_2 D_3$	4
		a_{12}	1 0000	$\bar{x}_2 x_3$	D_1	5
		a_{15}	1 0100	$\bar{x}_2 \bar{x}_3$	$D_1 D_3$	6
B_3	10	a_{12}	1 0000	1	D_1	7
B_4	11	a_{19}	1 1000	1	$D_1 D_2$	8

Table 6.11 The part of table of EMBer for Moore EFSM $P_{C1}Y_{CE}(\Gamma_{10})$

$K(a_m) \tau_1 \tau_2 \tau_3 T_1 T_2$	$Y(a_m) y_0 \dots y_5$	$K(B_i) z_1 z_2$	h	m
010 00	10 0100	00	9	8
010 01	10 1001	00	10	9
010 10	01 0011	01	11	10
010 11	00 0000	00	12	*
011 00	00 0010	10	13	11
011 01	00 0000	00	14	*
011 10	00 0000	01	15	*
011 11	00 0000	00	16	*

Let us discuss an example of synthesis for Moore EFSM $P_{C2}Y_{CE}(\Gamma_{10})$. The design method is the same as for Moore EFSM $P_{C2}Y_E$. Let it be $Z^1 = \{z_1\}$ and $Z^2 = \{z_2\}$. Let us save all codes we have for the case of $P_{C1}Y_{CE}(\Gamma_{10})$. Let us consider how to get the table of LUTer2.

This table includes the columns α_g, $K(\alpha_g)$, B_i, $K(B_i)$, Z_g^2, g. The class $B_i \in \Pi_{CE}$ is placed in the row number g of the table if $\alpha_g \in B_i$. The column Z_g^2 includes the variables $z_r \in Z^2$ equal to 1 in the code $K(B_i)$ from the row number $g (g = 1, \ldots, G1)$. The table of LUTer2 includes $G1 = 7$ rows in the discussed case (Table 6.12).

The sign * means that the corresponding chain code can be used for minimizing functions $z_r \in Z^2$. These functions are represented as

$$Z^2 = Z^2(\tau). \tag{6.34}$$

In the discussed case, the following equation can be derived from Table 6.12:

$$z_2 = \tau_1 \vee \bar{\tau}_2 \tau_3 \vee \tau_2 \bar{\tau}_3. \tag{6.35}$$

Table 6.12 Table of LUTer2 for Moore EFSM $P_{C2}Y_{CE}(\Gamma_{10})$

α_g	$K(\alpha_g)$	B_i	$K(B_i)$	Z_g^2	g
α_1	000	B_1	00	–	1
α_2	001	B_2	01	z_2	2
α_3	010	B_2	01	z_2	3
α_4	011	B_3	10	–	4
α_5	100	B_4	11	z_2	5
α_6	101	B_4	11	z_2	6
α_7	110	B_5	*	*	7

Fig. 6.13 Optimal chain codes for Moore EFSM $P_{C2}Y_{CE}(\Gamma_{10})$

$t_1 \backslash t_2 t_3$	00	01	11	10
0	α_1	α_2	α_5	α_4
1	*	α_3	α_6	α_7

Let us pint out that functions (6.34) can be simplified. Let us encode the chains $\alpha_g \in C_E(\Gamma_{10})$ as it is shown in Fig. 6.13.

Using these codes, the following equation can be found:

$$z_2 = \tau_3. \tag{6.36}$$

In this case, the block LUTer2 is absent.

6.3 Design of Moore EFSMs with Two Sources of Codes

The approach discussed in this section is based on ideas from Sect. 4.1. Let us adjust these ideas for ELCS-based Moore FSMs. Let us start from the model of PY_E Moore FSM (Fig. 6.1).

Let us form the sets C_E and Π_{CE} for some GSA Γ. Let us execute the optimal state assignment. Let $B_i \in \Pi_{CT}$ if the code $K(B_i)$ is represented by a single interval of R-dimensional Boolean space. If it is not true, that $B_i \in \Pi_{TC}$. The codes $K(B_i)$ should be generated by the block of transformer BTC if $B_i \in \Pi_{TC}$. It is enough R_{TC} variables (see (4.3)) to encode the classes $B_i \in \Pi_{TC}$.

Using (4.4)–(4.6), the number t_{BMO} can be found. It is equal to the number of unused outputs of EMBs from the EMBer. Let the following condition take place:

$$t_{BMO} = R_{TC} + 2. \tag{6.37}$$

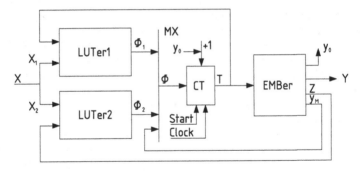

Fig. 6.14 Structural diagram of FPGA-based PY$_{E1}$ Moore EFSM

Table 6.13 Characteristics of GSA Γ_{11}

B_i	B_1	B_2		B_3		B_4			B_5			B_6		B_7	$C(a_m)$
α_g	α_1	α_2	α_3	α_4	α_5	α_6	α_7	α_8	α_9	α_{10}	α_{11}	α_{12}	α_{13}	α_{14}	
a_m	a_1	a_3	a_6	a_7	a_9	a_{11}	a_{14}	a_{16}	a_{18}	a_{20}	a_{23}	a_{25}	a_{27}	a_{29}	00
	a_2	a_4	–	a_8	a_{10}	a_{12}	a_{15}	a_{17}	a_{19}	a_{21}	a_{24}	a_{26}	a_{28}	a_{30}	01
	–	a_5	–	–	–	a_{13}	–	–	–	a_{22}	–	–	–	–	10
$K(\alpha_g)$	0000	0001	0010	0011	0100	0101	0110	0111	1000	1001	1010	1011	1100	1101	–
$K(B_i)$	000	001		010		011			100			101		110	–

One output of EMBer is used for generating y_0 and one y_M (see Fig. 4.1). If (6.37) takes place, then the model of PY$_{E1}$ Moore EFSM can be used (Fig. 6.14).

In this model, the block of LUTer1 implements the system of input memory functions (4.8), the block LUTer2 the system (4.9). The variables $z_r \in Z$ encode the classes $B_i \in \Pi_{TC}$, where $|Z| = R_{TC}$. As in the case of PY$_2$ Moore FSM, there is $X^1 \cup X^2 = X$. The variable y_M is used for control of the multiplexer MX.

The design method for PY$_{E1}$ includes the following steps:

1. Creating the set A.
2. Constructing the set of ELCS C_E.
3. Constructing the partition Π_{CE}.
4. Executing the optimal state assignment.
5. Finding the sets Π_{CT} and Π_{TC}.
6. Encoding the classes $B_i \in \Pi_{TC}$.
7. Constructing the table of LUTer1.
8. Constructing the table of LUTer2.
9. Constructing the table of EMBer.
10. Implementing the EFSM logic circuit.

Let us discuss an example of synthesis for Moore FSM PY$_{E1}(\Gamma_{11})$. The GSA Γ_{11} is rather complex. Because of it, we just show its characteristics in Table 6.13.

T_4T_5 \ $T_1T_2T_3$	000	001	011	010	110	111	101	100
00	a_1	a_5	a_{12}	a_8	a_{24}	a_{28}	a_{20}	a_{16}
01	a_2	a_6	a_{13}	a_9	a_{25}	a_{29}	a_{21}	a_{17}
11	a_4	a_7	a_{15}	a_{11}	a_{27}	*	a_{23}	a_{19}
10	a_3	*	a_{14}	a_{10}	a_{26}	a_{30}	a_{22}	a_{18}

Fig. 6.15 Optimal state codes for Moore EFSM $PY_{E1}(\Gamma_{11})$

The following sets and their parameters can be derived from Table 6.13: $A = \{a_1, \ldots, a_{30}\}$, $M = 30$; $C_E = \{\alpha_1, \ldots, \alpha_{14}\}$, $G1 = 14$, $\alpha_1 = \langle a_1, a_2 \rangle$, $\alpha_2 = \langle a_3, a_4, a_5 \rangle, \ldots, \alpha_{14} = \langle a_{25}, a_{30} \rangle$; $\Pi_{CE} = \{B_1, \ldots, B_7\}$, $I_{CE} = 7$, $B_1 = \{\alpha_1\}$, $B_2 = \{\alpha_2, \alpha_3\}, \ldots, B_7 = \{\alpha_{14}\}$. So, the steps 1–3 are already executed.

Let us encode the states $a_m \in A$ as it is shown in Fig. 6.15. Obviously, the condition (3.26) is satisfied for all ELCS $\alpha_g \in C_E(\Gamma_{11})$.

Let us define the code $K(B_i)$ using the state codes $K(a_m)$ where $a_m \in A(\alpha_g)$ and $\alpha_g \in B_i$. The following class codes can be found from Fig. 6.15: $K(B_1)=00001$, $K(B_2)=0010*$, $K(B_3)=010**$, $K(B_7)=11110$. Other classes are represented using more than a single generalized interval.

Now, the following sets can be found: $\Pi_{CT} = \{B_1, B_2, B_3, B_7\}$ and $\Pi_{TC} = \{B_4, B_5, B_6\}$. It gives $I_{TC} = 3$. Using (4.3), the following value can be found: $R_{TC} = 2$. It gives the set $Z = \{z_1, z_2\}$. Let it be $t_F = 16$ and $N = 5$. It gives the value $t_{BMO} = 6$. So, the condition (6.37) takes place and the model $PY_{E1}(\Gamma_{11})$ should be applied. Let us encode the classes $B_i \in \Pi_{TC}$ in the following way: $K(B_4) = 00$, $K(B_5) = 01$ and $K(B_6) = 10$.

Let the GSA Γ_{11} is characterized by the following system of GFT:

$$
\begin{aligned}
B_1 &\rightarrow x_1 a_3 \vee \bar{x}_1 a_6; \\
B_2 &\rightarrow x_2 a_7 \vee \bar{x}_2 a_9; \\
B_3 &\rightarrow x_1 a_{11} \vee \bar{x}_1 a_3 a_{14} \vee \bar{x}_1 \bar{x}_3 a_{16}; \\
B_4 &\rightarrow x_3 x_4 a_{11} \vee \bar{x}_1 \bar{x}_4 a_{18} \vee \bar{x}_3 x_5 a_{20} \vee \bar{x}_3 \bar{x}_5 a_{23}; \quad\quad (6.38) \\
B_5 &\rightarrow x_6 a_{25} \vee \bar{x}_6 a_{27}; \\
B_6 &\rightarrow a_{29}; \\
B_7 &\rightarrow a_1.
\end{aligned}
$$

The table of LUTer1 includes the following columns B_i, $K(B_i)$, $K(a_s)$, X_h, Φ_h, h (the same is true for the table of LUTer2). In the discussed case, the LUTer1 is represented by Table 6.14.

The state codes are taken from Fig. 6.15. Table 6.14 includes transitions for the classes $B_i \in \Pi_{CT}$. It is constructed using the GFTs for class B_1, B_2, B_3 (6.37).

Table 6.14 Table of LUTer1 of Moore EFSM $PY_{E1}(\Gamma_{11})$

B_i	$K(B_i)$	a_s	$K(a_s)$	X_h	Φ_h	h
B_1	0 0001	a_3	0 0010	x_1	D_4	1
		a_6	0 0101	$\bar{x}_1$	$D_3 D_5$	2
B_2	0 010*	a_7	0 0111	x_2	$D_3 D_4 D_5$	3
		a_9	0 1001	x_2	$D_2 D_5$	4
B_3	0 10**	a_{11}	0 1011	x_1	$D_2 D_4 D_5$	5
		a_{14}	0 1110	$\bar{x}_1 x_3$	$D_2 D_3$	6
		a_{16}	0 1001	$\bar{x}_1 \bar{x}_3$	D_1	7

Table 6.15 Table of LUTer2 of Moore EFSM $PY_{E1}(\Gamma_{11})$

B_i	$K(B_i)$	a_s	$K(a_s)$	X_h	Φ_h	h
B_4	00	a_{11}	0 1011	$x_3 x_4$	$D_2 D_4 D_5$	1
		a_{18}	1 0010	$x_3 \bar{x}_4$	$D_1 D_4$	2
		a_{20}	1 0100	$\bar{x}_3 \bar{x}_5$	$D_1 D_3$	3
		a_{23}	1 0111	$\bar{x}_3 \bar{x}_5$	$D_1 D_3 D_4 D_5$	4
B_5	01	a_{25}	1 1001	x_6	$D_1 D_2 D_5$	5
		a_{25}	1 1011	$\bar{x}_6$	$D_1 D_2 D_4 D_5$	6
B_6	10	a_{29}	1 1101	1	$D_1 D_2 D_3 D_5$	7

The table of LUTer1 is used for deriving the system (4.8). In the discussed case, for example, the following equations can be derived:

$$D_1 \to \bar{T}_1 T_2 \bar{T}_3 \bar{x}_1 \bar{x}_3;$$
$$D_5 \to \bar{T}_1 \bar{T}_2 \bar{T}_3 \bar{T}_4 T_5 \bar{x}_1 \vee \bar{T}_1 \bar{T}_2 T_3 \bar{T}_4 \vee \bar{T}_1 T_2 \bar{T}_3 x_1. \tag{6.39}$$

Table 6.15 represents the table of LUTer2 for the Moore EFSM $P_{E1} Y_E(\Gamma_{11})$. It is constructed using GFTs for the classes B_4, B_5, B_6 (5.37).

The table of LUTer2 is used for deriving the system (4.9). In the discussed case, for example, the following equations can be derived:

$$D_1 \to \bar{z}_1 \bar{z}_2 x_3 \bar{x}_4 \vee \bar{z}_1 \bar{z}_2 \bar{x}_3 \vee \bar{z}_1 z_2 \vee z_1 \bar{z}_2;$$
$$D_5 \to \bar{z}_1 \bar{z}_2 x_3 \bar{x}_4 \vee \bar{z}_1 \bar{z}_2 \bar{x}_3 x_5 \vee \bar{z}_1 z_2 \vee z_1 \bar{z}_2. \tag{6.40}$$

Let us use $y_M = 0(y_M = 1)$ to indicate that the input memory functions $\Phi_1(\Phi_2)$ should be loaded into CT. Using this rule and class codes for $B_i \in \Pi_{TC}$, the table of EMBer can be constructed. It includes the columns $K(a_m)$, $Y(a_m)$, y_0, y_M, $K(B_i)$, m. If $(a_m = O_g \& \alpha_g \in B_i \& B_i \in \Pi_{TC}) = 1$ then the code $K(B_i)$ is placed in the m-th row of the table together with $y_M = 1$ If $a_m \neq O_g$, then the variable y_0 is placed in the row number m of the table. This table is constructed in the trivial way. We do not discuss it in this Chapter.

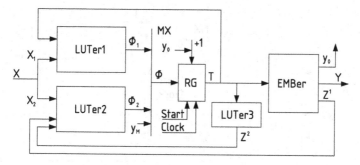

Fig. 6.16 Structural diagram of FPGA-based PY_{E2} Moore EFSM

If $t_{BMO} > 0$ but the condition (6.37) is violated, then only R_{TC1} bits can be generated by the EMBer:

$$R_{TC1} = R_{TC} - t_{TMO}. \tag{6.41}$$

The rest of bits is generated by the block LUTer3. It leads to the Moore EFSM PY_{E2} (Fig. 6.16).

This model combines the features of PY_{E1} and $P_{C2}Y_E$ EFSMs. The same is true for the corresponding design methods. We leave this EFSM to our reader.

Now, let us discuss application of this idea for Moore EFSMs with code sharing. Let us start from the model of PY_{EC} Moore EFSM (Fig. 6.8). In this case, there is no need in the optimal state assignment. We should execute the optimal chain assignment.

Let us find the sets A, C_E and Π_{CE} for a given GSA Γ. Let us execute the natural state assignment (3.65). Let us encode the chains $\alpha_g \in C_E$ in the optimal way. Let us represent the set Π_{CE} as $\Pi_{CE} = \Pi_{RG} \cup \Pi_{TC}$. There is the relation $B_i \in \Pi_{RG}$ if the code $K(B_i)$ is represented by a single interval of R_{G1}-dimensional Boolean space. Otherwise, there is the relation $B_i \in \Pi_{TC}$. The chain codes should be transformed if $(\alpha_g \in B_i) \& B_i \in \Pi_{TC}) = 1$. It is enough R_{TC} variables $z_r \in Z$ to encode the classes $B_i \in \Pi_{TC}$. Let the condition (6.37) take place. In this case, the following structural diagram (Fig. 6.17) is proposed for PY_{EC1} Moore EFSM.

In this case, the LUTer1 generates the input memory functions

$$\Phi_1 = \Phi_1(\tau, X_1). \tag{6.42}$$

The LUTer2 implements the input memory functions

$$\Phi_2 = \Phi_2(\tau, X_2). \tag{6.43}$$

The following system of equations is implemented by the multiplexer MX:

$$\Phi = y_M \Phi_1 \vee \bar{y}_M \Phi_2. \tag{6.44}$$

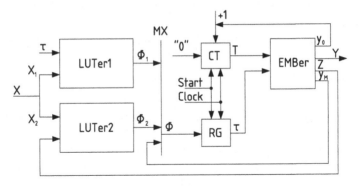

Fig. 6.17 Structural diagram of FPGA-based PY$_{EC1}$ Moore EFSM

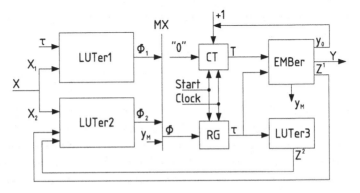

Fig. 6.18 Structural diagram of FPGA-based PY$_{EC2}$ Moore EFSM

The EMBer implements functions y_0, Y, Z and y_M depended on variables from the sets τ and T.

The proposed design method includes the following steps for PY$_{EC1}$ Moore EFSM:

1. Creating the set A.
2. Constructing the set of ELCS C_E.
3. Constructing the partition Π_{CE}.
4. Executing the natural state assignment.
5. Executing the optimal chain assignment.
6. Finding the sets Π_{RG} and Π_{TC}.
7. Encoding of the classes $B_i \in \Pi_{TC}$.
8. Constructing the table of LUTer1.
9. Constructing the table of LUTer2.
10. Constructing the table of EMBer2.
11. Implementing the EFSM logic circuit.

If the condition (6.37) is violated, then R_{TC1} functions $z_r \in Z$ belong to the set Z^2. It leads to PY$_{EC2}$ Moore EFSM (Fig. 6.18).

Fig. 6.19 Structural
diagram of HFPGA-based
PY_E Moore EFSM

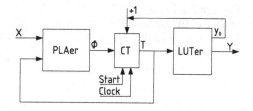

The block LUTer3 generates the functions

$$Z^2 = Z^2(\tau). \tag{6.45}$$

The design methods are practically the same for PY_{CE1} and PY_{CE2} EFSMs. But it is necessary one more step for design of PY_{EC2} EFSM. It is connected with constructing the table of LUTer3. We do not discuss these methods in our book.

6.4 Design of Moore EFSMs with HFPGAs

There are no embedded memory blocks in modern hybrid FPGAa [8, 9]. So, only PLA macrocells and LUTs can be used for implementing the logic circuits of FSMs. Let the symbol P_H means that HFPGAs are used for implementing an EFSM logic circuit. The structural diagram (Fig. 6.19) represents the $P_H Y_E$ Moore EFSM.

In $P_H Y_E$ FSM, the PLAer implements the system of input memory functions $\Phi = \Phi(T, X)$, whereas the LUTer generates microoperations $Y = Y(T)$ and $y_0 = y_0(T)$. Let the symbol $PLA(s_p, t_p, q_t)$ denotes a PLA macrocell having s_p inputs, t_p outputs and q_p product terms. Let the following conditions take places:

$$s_p \geq L + R; \tag{6.46}$$

$$t_p \geq R; \tag{6.47}$$

$$q_p \geq H_E. \tag{6.48}$$

In this case, it is necessary only a single PLA for implementing the circuit of PLAer. If the condition (6.47) is violated, then the "expansion of PLA outputs" should be executed. If the condition (6.48) is violated, then the "expansion of PLA terms" should be executed [3]. If the condition (6.46) is violated, then different methods from [1] should be used. We do not discuss these cases in our book.

Let a LUT in use have s_L inputs. Let the following condition take place:

$$s_L \geq R. \tag{6.49}$$

In this case, only $N + 1$ LUTs are used in the LUTer. If the condition (6.49) is violated, the different methods of functional decomposition [12] should be used.

Fig. 6.20 Structural
diagram of HFPGA-based
$P_{HC}Y_E$ Moore EFSM

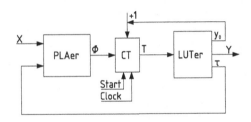

Let us discuss some approaches leading to diminishing the numbers of PLA cells and LUT elements in logic circuits of $P_H Y_E$ EFSMs. In the condition (6.48) is violated, then the method of optimal state assignment can be used. It leads to $P_{H0}Y_E$ Moore EFSM having the same structural diagram as the one shown in Fig. 6.20. In this method does not give the minimum possible amount of terms, then the transformation of state codes can be used. It leads to $P_{HC}Y_E$ Moore EFSM (Fig. 6.20).

The design methods are obvious for $P_H Y_E$, $P_{H0}Y_E$ and $P_{HC}Y_E$ EFSMs. Because of it, we do not discuss them. Let us discuss the situation when the following condition takes place:

$$t_p > R. \tag{6.50}$$

Let us find the value of t_A:

$$t_A > t_p - R. \tag{6.51}$$

Let it be q_A unused terms in PLAer:

$$q_A = q_p - H(\Phi). \tag{6.52}$$

In (6.52), $H(\Phi)$ is the number of terms in the system (2.9).

Let us find the equations for functions (2.16) and y_0. Let us divide the set $y_0 \cup Y$ by two subsets Y^1 and Y^2. Let the set Y^1 satisfy to the following conditions:

$$N(Y^1) \leq t_A; \tag{6.53}$$

$$H(Y^1) \leq q_A.i \tag{6.54}$$

In (6.53)–(6.54), $N(Y^1)$ is the number of elements in the set Y^1, whereas $H(Y^1)$ is the number of terms in the functions $y_n \in Y^1$. Let us point out that it is quite possible the following relation: $y_0 \in Y^1$. Let $Y^1 \neq \emptyset$, then the functions from Y^1 can be implemented by PLAer. It results in $P_{H1}Y_E$ Moore EFSM (Fig. 6.21).

Obviously, the $P_{H01}Y_E$ Moore EFSM has the same structure. Let us discuss an example of design for $P_{H01}Y_E(\Gamma_{10})$. The proposed design metod includes the following steps:

1. Creating the set of states A.
2. Constructing the set of ELCS C_E.
3. Constructing the partition Π_{CE}.

Fig. 6.21 Structural
diagram of HFPGA-based
$P_{HC}Y_E$ Moore EFSM

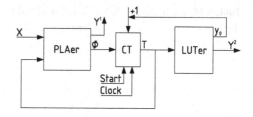

4. Executing the optimal natural state assignment.
5. Creating the preliminary table of PLAer.
6. Dividing the set $Y \cup y_0$ by subsets Y^1 and Y^2.
7. Creating the final table of PLAer.
8. Creating the table of LUTer.
9. Implementing the FSM logic circuit.

The sets A, C_E and Π_{CE} are already found for the Moore EFSM $PY_E(\Gamma_{10})$. They can be derived from Table 6.6. Let us encode the states $a_m \in A$ as it is shown in Fig. 6.3. The preliminary table of PLAer is the same as the table of LUTer (Table 6.1). The Eqs. (6.4)–(6.6) represent the system Φ. Let us rewrite them in the following system:

$$D_1 = F_1 \vee F_2 \vee F_3;$$
$$D_2 = F_4 \vee F_5 \vee F_6; \qquad (6.55)$$
$$D_3 = F_7 \vee F_8 \vee F_3.$$

The value $H_E(\Phi) = 8$ can be derived from the system (6.55). Let the HFPGA chip in use have macrocells PLA with $s_p = 12$, $t_p = 6$ and $q_p = 17$. Using (6.52), it can be found $q_A = 16 - 8 = 8$. Using (6.51), it can be found the value $t_A = 6 - 3 = 3$. So, it is possible to generate up to 3 functions $y_n \in Y$ by PLAer.

Let us form the system (2.16) for the discussed case. Using Fig. 6.2, the following system can be found:

$$y_1 = A_2 \vee A_4 \vee A_7 \vee A_{10} \vee A_{12} \vee A_{14} \vee A_{17} \vee A_{19};$$
$$y_2 = A_2 \vee A_2 \vee A_6 \vee A_7 \vee A_9 \vee A_{12} \vee A_{15} \vee A_{18};$$
$$y_3 = A_3 \vee A_5 \vee A_8 \vee A_{15} \vee A_{17}; \qquad (6.56)$$
$$y_4 = A_4 \vee A_{10} \vee A_{11} \vee A_{12} \vee A_{14};$$
$$y_5 = A_6 \vee A_9 \vee A_{10} \vee A_{13} \vee A_{16} \vee A_{18}.$$

After minimizing the system (6.56), it can be found that the function y_1 includes $H(y_1) = 4$ terms. Also, the following values can be found $H(y_2) = 6$, $H(y_3) = 4$, $H(y_4) = 3$ and $H(y_5) = 2$. So, only a pair of microoperations can be chosen for implementing by the PLAer. Let su choose the functions y_4 and y_5. They are represented by the following minimized equations:

Table 6.16 Final table of PLAer of Moore EFSM $P_{H01}Y_E(\Gamma_{10})$

Inputs $T_1 \ldots T_5 x_1 \ldots x_4$	Terms	Outputs $D_1 D_2 D_3 y4 y5$
0*1** *1**	1	100 00
0*1** **0*	2	100 00
*10** ****	3	101 00
000** 0***	4	010 00
0*1** *0**	5	010 00
100** ****	6	010 00
000** ****	7	001 00
0*1** *1*1	8	001 00
*0101 ****	9	000 10
110*0 ****	10	000 10
*1010 ****	11	000 10
*1**1 ****	12	000 01
110 **	13	000 01

$$y_4 = \bar{T}_2 T_3 \vee T_4 T_5 \vee T_1 T_2 \bar{T}_3 \bar{T}_5 \vee T_2 \bar{T}_3 T_4 \bar{T}_5 = F_9 \vee F_{10} \vee F_{11};$$
$$y_5 = T_2 T_5 \vee T_3 T_4 \bar{T}_5 = F_{12} \vee F_{13}. \tag{6.57}$$

So, now there are two sets: $Y^1 = \{y_4, y_5\}$ and $Y^2 = \{y_0, y_1, y_2, y_3\}$. Of course, the pair y_2, y_5 could be taken, as well as the pair y_3, y_4.

The final table of PLAer includes the columns *Inputs*, *Terms* and *Outputs*. It is just a table using for programming a PLA [1]. In the discussed case, it includes 13 rows (Table 6.16). If some term includes the direct value of some variable, it corresponds to 1. If some term includes the complement value of some variable, it corresponds to 0. If there is no variable in the term, then it corresponds to *. If some function depends on a term, it is denoted as 1. Otherwise, it is denoted as 0.

We hope that the connection is obvious between the systems (6.4)–(6.6), (6.53), (6.57) and Table 6.16. The column *Terms* includes the numbers of terms of systems (6.53) and (6.57).

Let it be $s_L = 5$ in the discussed case. So, the condition (6.49) is true. So, only 4 LUTs are necessary for implementing the set Y^2. The logic circuit of Moore EFSM $P_{H01}Y_E(\Gamma_{10})$ is shown in Fig. 6.22.

Let us find out that I_{TC} classes $B_i \in \Pi_{CE}$ cannot be represented by single intervals of R-dimensional Boolean space. It means that the set Π_{CE} is represented as $\Pi_{CT} \cup \Pi_{TC}$. Let us encode the classes $B_i \in \Pi_{TC}$ by binary codes $K(B_i)$ having R_{TC} bits. The value of R_{TC} is determined by (4.3). Let us use the variables $z_r \in Z$ for encoding of the classes $B_i \in \Pi_{CE}$. In this case, the model of $P_H Y_{E1}$ Moore EFSM (Fig. 6.23) is proposed.

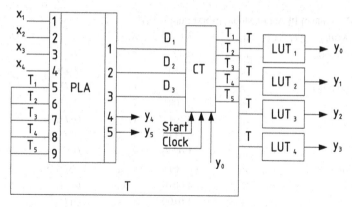

Fig. 6.22 Logic circuit of Moore EFSM $P_{H01}Y_E(\Gamma_{10})$

Fig. 6.23 Structural diagram of P_HY_{E1} Moore EFSM

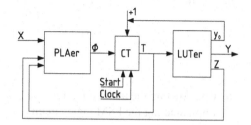

In this model, the PLAer implements the system

$$\Phi = \Phi(T, Z, X). \tag{6.58}$$

The LUTer implements functions (2.16), (3.29) and

$$Z = Z(T). \tag{6.59}$$

This model can be applied if the following condition takes place:

$$S_p \geq L + R + R_{TC}. \tag{6.60}$$

It is known that PLA macrocells have the wide fan-in which is equal up to 30 [8, 9]. It means that the condition (6.60) takes place in many practical cases.

There are the following steps in the proposed design method targeting P_HY_{E1} Moore EFSM:

1. Creating the set of states A, set of ELCS C_E and the partition Π_{CE}
2. Executing the optimal natural state assignment.
3. Finding the sets Π_{CT} and Π_{TC}.
4. Encoding of the classes $B_i \in \Pi_{TC}$.
5. Constructing the table of PLAer.

Table 6.17 Table of PLAer of Moore EFSM $P_{HE1}Y(\Gamma_{11})$

B_i	$K(B_i)z_1z_2T_1\ldots T_5$	a_s	$K(a_s)$	X_h	Φ_h	h
B_1	00 00001	a_3	0 0010	x_1	D_4	1
		a_6	0 0101	$\bar{x}_1$	D_3D_5	2
B_2	00 0010*	a_7	0 0111	x_2	$D_3D_4D_5$	3
		a_9	0 1001	$\bar{x}_2$	D_2D_5	4
B_3	00 010**	a_{11}	0 1011	x_1	$D_2D_4D_5$	5
		a_{14}	0 1110	$\bar{x}_1x_3$	$D_2D_3D_4$	6
		a_{16}	1 0000	$\bar{x}_1\bar{x}_3$	D_1	7
B_4	01 *****	a_{11}	0 1011	x_3x_4	$D_2D_4D_5$	8
		a_{18}	1 0010	$x_3\bar{x}_4$	D_1D_4	9
		a_{20}	1 0100	$\bar{x}_3x_5$	D_1D_3	10
		a_{23}	1 0111	$\bar{x}_3\bar{x}_5$	$D_1D_2D_4D_5$	11
B_5	10 *****	a_{25}	1 1001	x_6	$D_1D_2D_5$	12
		a_{27}	1 1011	$\bar{x}_6$	$D_1D_2D_4D_5$	13
B_6	11 *****	a_{29}	1 1101	1	$D_1D_2D_3D_5$	14

6. Constructing the table of LUTer.
7. Implementing the FSM logic circuit.

Let us discuss an example of design for Moore EFSM $P_{HE1}Y(\Gamma_{11})$. The characteristics of GSA Γ_{11} are shown in Table 6.12. The optimal natural state codes are shown in Fig. 6.15. They are the following: $K(B_1) = 00001$, $K(B_2) = 0010*$, $K(B_3) = 010**$, $K(B_7) = 11110$. There is the set $\Pi_{TC} = \{B_4, B_5, B_6\}$ having $I_{TC} = 3$. The number of R_{TC} is determined as

$$R_{TC} = \lceil \log_2(I_{TC} + 1)\rceil. \tag{6.61}$$

The value 1 is added to I_{TC} to take into account the relation $B_i \in \Pi_{CT}$. It should be represented by an unique code using the variables $z_r \in Z$.

Let us encode the classes $B_i \in \Pi_{TC}$ in the following manner: $K(B_4) = 01$, $K(B_5) = 10$ and $K(B_6) = 11$. Let us use the code 00 to show that $B_i \notin \Pi_{TC}$. The table of LUTer is represented by Table 6.17 for a given example.

The table of PLAer is used for deriving the functions (6.58). These functions depend on the following terms:

$$F_h = \left(\bigwedge_{r=1}^{R_{TC}} z_r^{l_{ir}}\right)\left(\bigwedge_{r=1}^{R} T_r^{l_{ir}}\right)X_h. \tag{6.62}$$

For example, the following terms (6.62) can be obtained from Table 6.17: $F_1 = \bar{z}_1\bar{z}_2\bar{T}_1\bar{T}_2\bar{T}_3\bar{T}_4T_5$, $F_6 = \bar{z}_1\bar{z}_2\bar{T}_1T_2\bar{T}_3$, $F_{12} = z_1\bar{z}_2x_6$. The table of LUTer includes the columns $K(a_m)$, $Y(a_m)$, $Z(a_m)$, m. The column $Z(a_m)$ includes the codes $K(B_i)$

Fig. 6.24 Structural diagram
of $P_{H1}Y_{EC1}$ Moore FSM

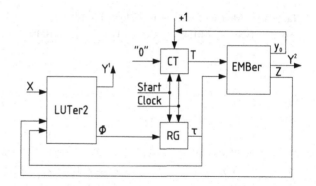

for the classes $B_i \in \Pi_{TC}$. This table corresponds to $R_{TC} + N + 1$ truth tables. Each table corresponds to as single function from the set $y_0 \cup Y \cup Z$.

Let us point out that the $P_H Y_{E1}$ EFSM is based on the results from [5, 6]. These ideas can be applied for the case of $P_{H1} Y_{E1}$ EFSM. Its design method is combined from the methods for $P_{H1} Y_E$ and $P_H Y_{E1}$ Moore EFSMs. The same approach can be used for EFSMs with code sharing. Let us, for example, discuss the design method for EFSM $P_{H1} Y_{EC1}$. Its structural diagram is shown in Fig. 6.24.

In this model, the PLAer implements input memory functions:

$$\Phi = \Phi(\tau, Z, X). \tag{6.63}$$

The LUTer implements the functions

$$y_0 = y_0(\tau, T); \tag{6.64}$$
$$Z = Z(\tau, T); \tag{6.65}$$
$$Y^2 = Y^2(\tau, T). \tag{6.66}$$

Also, the PLAer implements some subset $Y^1 \subset Y$, which is represented as

$$Y^1 = Y^1(\tau, Z, X). \tag{6.67}$$

The design method includes the following steps for this model:

1. Creating the set of states A.
2. Constructing the set of ELCS C_E.
3. Constructing the set Π_{CE}.
4. Optimal encoding of the ELCS $\alpha_g \in C_E$.
5. Natural encoding of the states $a_m \in A(\alpha_g)$.
6. Encoding of the classes $B_i \in \Pi_{TC}$
7. Finding the sets Y^1 and Y^2.
8. Constructing the table of PLAer.

Table 6.18 Models of ELCS-based Moore FSMs

No	Type	BIMF	BTC	BMO	Basis	Comments
1.	PY_E	$\Phi = \Phi(T, X)$	–	$y_0 = y_0(T)$	FPGA	Base structure
2.	P_0Y_E			$Y = Y(T)$		
3.	$P_{C1}Y_E$	$\Phi = \Phi(\tau, X)$	–	$\tau = \tau(T)$		
4.	$P_{C2}Y_E$		$\tau^2 = \tau^2(T)$	$\tau^1 = \tau^1(T)$		
5.	PY_{EC}	$\Psi = \Psi(\tau, X)$	–	$y_0 = y_0(\tau, T)$		Code sharing
6.	P_0Y_{EC}			$Y = Y(\tau, T)$		
7.	$P_{C1}Y_{EC}$	$\Psi = \Psi(Z, X)$		$Z = Z(\tau, T)$		
8.	$P_{C2}Y_{EC}$		$Z^2 = Z^2(\tau)$	$Z^1 = Z^1(\tau, T)$		
8.	$P_{C2}Y_{EC}$		$Z^2 = Z^2(\tau)$	$Z^1 = Z^1(\tau, T)$		
9.	PY_{E1}	$\Phi_1 = \Phi_1(T, X)$ $\Phi_2 = \Phi_2(Z, X)$	–	$y_0 = y_0(T)$ $Y = Y(T)$ $y_M = y_M(T)$ $Z = Z(T)$		Two sources of codes for base structure
10.	$P_{C2}Y_{E2}$		$Z^2 = Z^2(\tau)$	$Z^1 = Z^1(T)$		
11.	PY_{E2}	$\Psi_1 = \Psi_1(\tau, X_1)$ $\Phi_2 = \Phi_2(Z, X)$	–	$y_0 = y_0(\tau, T)$ $Y = Y(\tau, T)$ $y_M = y_M(\tau, T)$ $Z = Z(\tau, T)$		Two sources for code sharing
12.	PY_{EC2}		$Z^2 = Z^2(\tau)$	$Z^1 = Z^1(\tau, T)$		
13.	P_HY_E	$\Phi = \Phi(T, X)$	–	$y_0 = y_0(\tau, T)$	HFPGA	Base structure
14.	$P_{H0}Y_{EC}$			$Y = Y(T)$		
15.	$P_{HC}Y_E$	$\Phi = \Phi(\tau, X)$		$\tau = \tau(T)$		
16.	$P_{H1}Y_E$	$Y^1 = Y^1(T)$		$Y^2 = Y^2(T)$		
17.	$P_{H01}Y_E$					
18.	$P_{HC1}Y_E$					
19.	P_HY_{E1}	$\Phi = \Phi(T, Z, X)$		$y_0 = y_0(T)$ $Y = Y(T)$ $Z = Z(T)$		Two sources
20.	$P_{H1}Y_{E1}$	$Y^1 = Y^1(T, Z)$		$Y^2 = Y^2(T)$		
21.	$P_{H1}Y_{EC1}$	$\Psi = \Psi(\tau, Z, X)$ $Y^1 = Y^1(\tau, Z)$		$y_0 = y_0(\tau, T)$ $Y^2 = Y^2(\tau, T)$ $Z = Z(\tau, T)$		Code sharing with two sources

9. Constructing the table of LUTer.
10. Implementing the FSM logic circuit.

This method combines steps form different methods discussed in this Chapter. We hope that a reader will not have troubles with designing the logic circuit of $P_{H1}Y_{EC1}$ Moore EFSM.

We show all discussed models of ELCS-based Moore FSMs in Table 6.18. More structures can be added for the case of HFPGA-based EFSMs with code sharing:

1. The model $P_H Y_{EC}$ is a base model with code sharing. It is the same as the model of PY_{EC} Moore EFSM (Fig. 6.8). Of course, in this case the BIMF is represented by the PLAer, whereas the BMO is represented by LUTer.
2. The model $P_{H0} Y_{EC}$ is a base model with the optimal chain encoding.
3. The model $P_{HC} Y_{EC}$ is a base model with the transformation of chain codes into class codes $K(B_i)$.

These three models are not included in Table 6.18. It is done because we did not discuss them in this Chapter. But we hope that our reader will be able to work out the corresponding design methods. Of course, they are based on the methods already discussed in this Chapter. Now let us discuss design methods targeting Moore FSMs based on the normal linear chains of states.

References

1. Adamski, M., Barkalov, A.: Architectural and Sequential Synthesis of Digital Devices. University of Zielona Góra Press, Zielona Góra (2006)
2. Baranov, S.: Logic and System Design of Digital Systems. TUT Press, Tallinn (2008)
3. Baranov, S.I.: Logic Synthesis for Control Automata. Kluwer Academic Publishers, Dordrecht (1994)
4. Barkalov, A., Titarenko, L.: Logic Synthesis for FSM-Based Control Units. Lecture Notes in Electrical Engineering, vol. 53. Springer, Berlin (2009)
5. Barkalov, A., Titarenko, L., Chmielewski, S.: Reduction in the number of PAL macrocells in the circuit of Moore FSM. Int. J. Appl. Math. Comput. Sci. **17**(4), 565–575 (2007)
6. Barkalov, A., Titarenko, L., Malchera, R., Soldatov, K.: Hardware reduction in FPGA-based Moore FSM. J. Circuits Syst. Comput. **22**(3), 1350006-1–1350006-20 (2013)
7. Barkalov, A., Węgrzyn, M.: Design of Control Units with Programmable Logic. UZ Press, University of Zielona Góra, Zielona Góra (2006)
8. Kaviani, A., Brown, S.: The hybrid field-programmable architecture. IEEE Des. Test Comput. **16**(2), 74–83 (1999)
9. Krishnamoorthy, S., Tessier, R.: Technology mapping algorithms for hybrid FPGAs containing lookup tables and PLAs. IEEE Trans. Comput. Aided Des. Integr. Circuits Syst. **22**(5), 545–559 (2003)
10. Markovitz, A.: Introduction to Logic Desgin. McGraw Hill, New York (2012)
11. Roth, C., Kinney, L.: Fundamentals of Logic Design. Cengage Learnig, New York (2009)
12. Scholl, C.: Functional Decomposition with Application to FPGA Synthesis. Kluwer Academic Publishers, Boston (2001)

Chapter 7
Hardware Reduction for Moore NFSMs

Abstract The Chapter is devoted to hardware reduction targeting the normal LCS-based Moore FSMs. Firstly, the optimization methods are proposed for the base model of NFSM. They are based on the executing either optimal state assignment or transformation of state codes. Two different models are proposed for the case of code transformation. They depend on the numbers of microoperations of FSM and outputs of EMB in use. The models are discussed based on the principle of code sharing. In this case, the state code is represented as a concatenation of the code of normal LCS and the code of component inside this chain. The last part of the chapter is devoted to design methods targeting the hybrid FPGAs.

7.1 Optimization of NFSMs with the Base Structure

The only difference exists between EFSMs and NFSMs. The NFSMs are based on natural linear chains of states. Because of it, we can use all optimization methods proposed in Sect. 6.1. Of course, the set $C_N = \{\beta_1, \ldots, \beta_{G2}\}$. should be found instead of the set C_E. So, the following four models can be generated:

1. The model of PY_N Moore NFSM has the same structural diagram as the one shown in Fig. 6.1.
2. The model of P_0Y_N Moore NFSM is based on the optimal state assignment. It has the same structural diagram as the one shown in Fig. 6.1. The partition $\Pi_{CN} = \{B_1, \ldots, B_{IN}\}$ should be constructed. Each class $B_i \in \Pi_{CN}$ includes pseudoequivalent NLCS $\beta \in C_N$. The optimal codes $K(a_m)$ should satisfy the following condition:

$$K(a_{gi+1}) = K(a_{gi}) + 1 \quad (g = 1, \ldots, G2). \tag{7.1}$$

3. The model of $P_{C1}Y_N$ Moore NFSM is based on the transformation of state codes $K(a_m)$ into the class codes $K(B_i)$. It has the same structural diagram as the one shown in Fig. 6.5. The set τ includes R_{CN} variables:

© Springer International Publishing AG 2018
A. Barkalov et al., *Logic Synthesis for Finite State Machines Based on Linear Chains of States*, Studies in Systems, Decision and Control 113,
DOI 10.1007/978-3-319-59837-6_7

$$R_{CN} = \lceil \log_2 IN \rceil. \tag{7.2}$$

This approach can be used if the following condition takes place:

$$t_F \geq N + 1 + R_{CN}. \tag{7.3}$$

In (7.3), the symbol t_F stands for the number of outputs of EMB. This number should provide the following condition:

$$V_0 \geq M. \tag{7.4}$$

In (7.4), the symbol V_0 stands for the number of cells of EMB for a given number of outputs t_F.

4. The model of $P_{C2}Y_N$ Moore NFSM is based on the same principle of code transformation. It has the same structure as the one shown in Fig. 6.6. It can be used if two conditions take place. The first of them is the condition (6.18). The second is the following condition:

$$t_F < N + R_{CN} + 1. \tag{7.5}$$

Let us discuss this model in details. The proposed synthesis method includes the following steps:

1. Creating the set of states A for a given GSA Γ.
2. Constructing the set of NLCSs C_N.
3. Constructing the partition $\Pi_{CN} = \{B_1, \ldots, B_{IN}\}$
4. Executing the natural state assignment (7.1).
5. Encoding of the classes $B_i \in \Pi_{CN}$.
6. Constructing the table of LUTer1.
7. Constructing the table of LUTer2.
8. Constructing the table of EMBer.
9. Implementing the FSM logic circuit.

Let us discuss an example of synthesis for Moore NFSM $P_{C2}Y_N(\Gamma_{12})$. The initial GSA Γ_{12} is shown in Fig. 7.1. It is marked by the states of Moore FSM using rules [1]. The following sets and their characteristics can be derived from Fig. 7.1: $A = \{a_1, \ldots, a_{16}\}$, $M = 16$, $X = \{x_1, \ldots, x_3\}$, $L = 5$, $Y = \{y_1, \ldots, y_7\}$, $N = 7$. Using (2.15), it can be found that $R = 4$. It gives the sets $T = \{T_1, \ldots, T_4\}$ and $\Phi = \{D_1, \ldots, D_4\}$.

Let us apply the procedure P_4 to GSA Γ_{12}. It gives the set of normal linear chains of states C_N having $G2 = 8$ elements. There are the following chains: $\beta_1 = \langle a_1, a_2 \rangle$, $\beta_2 = \langle a_3, \ldots, a_6 \rangle$, $\beta_3 = \langle a_7, a_8 \rangle$, $\beta_4 = \langle a_9, a_{10} \rangle$, $\beta_5 = \langle a_{11}, a_{12} \rangle$, $\beta_6 = \langle a_{13} \rangle$, $\beta_7 = \langle a_{14} \rangle$ and $\beta_8 = \langle a_{15}, a_{16} \rangle$.

It can be found the set $\Pi_{CN} = \{B_1, \ldots, B_4\}$ where $B_1 = \{\beta_1\}$, $B_2 = \{\beta_2, \beta_3, \beta_4\}$, $B_3 = \{\beta_5, \beta_6, \beta_7\}$ and $B_4 = \{\beta_8\}$. So, there is $R_{CN} = 2$. It gives the set $\tau = \{\tau_1, \tau_2\}$. The outcome of natural state assignment is shown in Fig. 7.2.

Fig. 7.1 Initial marked
GSA Γ_{12}

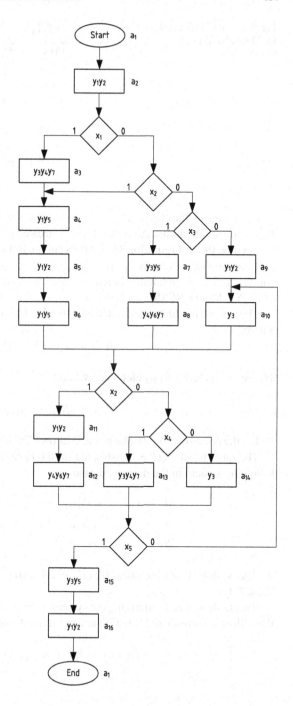

Fig. 7.2 Natural state codes
for Moore NFSM
$P_{C2}Y_N(\Gamma_{12})$

T_3T_4 \ T_1T_2	00	01	11	10
00	a_1	a_5	a_{13}	a_9
01	a_2	a_6	a_4	a_{10}
11	a_4	a_8	a_{16}	a_{12}
10	a_3	a_7	a_{15}	a_{11}

Let us encode the classes $B_i \in \Pi_{CN}$ in the trivial way: $K(B_1) = 00, \ldots, K(B_4) = 11$. Let us point out that the code 11 can be treated as "don't care" input assignment.

Let it be the configuration 16×9 (bits) in the FPGA chip in use. So, the condition (7.4) is satisfied. There is $t_F = 9$. There is $N + 1 + R_{CN} = 10$. It means that the condition (7.3) is violated, whereas the condition (6.18) takes place. So, the model of $P_{C2}Y_N$ Moore NFSM can be used.

The set τ should be derived using (6.17). The set τ^1 includes R_{CN1} elements where:

$$R_{CN1} = t_F - (N + 1). \tag{7.6}$$

The set τ^2 includes R_{CN2} elements, where:

$$R_{CN2} = R_{CN} - R_{CN1}. \tag{7.7}$$

Let the condition (6.22) take place. It allows finding sets: $\tau^1 = \{\tau_1\}$ and $\tau^2 = \{\tau_2\}$.

The table of LUTer1 is constructed on the base of the system GFT. There is the following system in the discussed case:

$$\begin{aligned}
B_1 &= x_1 a_3 \vee \bar{x}_1 x_2 a_4 \vee \bar{x}_1 \bar{x}_2 x_2 a_7 \vee \bar{x}_1 \bar{x}_2 \bar{x}_3 a_9; \\
B_2 &= x_2 a_{11} \vee \bar{x}_2 x_4 a_{13} \vee \bar{x}_2 \bar{x}_4 a_{14}; \\
B_3 &= x_5 a_{15} \vee \bar{x}_5 a_{10}.
\end{aligned} \tag{7.8}$$

The system (7.8) includes $H_N(\Gamma_{12}) = 9$ terms. It is the number of rows for Table 7.1.

This table is used for deriving the system of input memory functions. For example, the following minimized function can be derived from Table 7.1:

$$D_1 = \bar{\tau}_1 \bar{\tau}_2 \bar{x}_1 \bar{x}_2 \bar{x}_3 \vee \bar{\tau}_1 \tau_2 \vee \tau_1 \bar{\tau}_2. \tag{7.9}$$

Table 7.1 Table of LUTer1 for Moore NFSM $P_{C2}Y_N(\Gamma_{12})$

B_i	$K(B_i)$	a_s	$K(a_s)$	X_h	Φ_h	h
B_1	00	a_3	0010	x_1	D_3	1
		a_4	0011	$\bar{x}_1 x_2$	$D_3 D_4$	2
		a_7	0110	$\bar{x}_1 \bar{x}_2 x_3$	$D_2 D_3$	3
		a_9	1000	$\bar{x}_1 \bar{x}_2 \bar{x}_3$	D_1	4
B_2	01	a_{11}	1010	x_2	$D_1 D_3$	5
		a_{13}	1100	$\bar{x}_2 x_4$	$D_1 D_2$	6
		a_{14}	1101	$\bar{x}_2 \bar{x}_4$	$D_1 D_2 D_4$	7
B_3	10	a_{15}	1110	x_5	$D_1 D_2 D_3$	8
		a_{10}	1001	$\bar{x}_5$	$D_1 D_4$	9

Table 7.2 Table of LUTer2 for Moore NFSM $P_{C2}Y_N(\Gamma_{12})$

a_m	$K(a_m)$	B_i	$K(B_i)$	τ_m^2	m
a_2	0001	B_1	00	–	2
a_6	0101	B_2	01	τ_2	6
a_8	0111	B_2	01	τ_2	8
a_{10}	1001	B_2	01	τ_2	10
a_{12}	1011	B_3	10	–	12
a_{13}	1100	B_3	10	–	13
a_{14}	1101	B_3	10	–	14

The table of LUTer2 is constructed only for outputs of chains $\beta_g \in C_N$. It contains the columns a_m, $K(a_m)$, B_i, $K(B_i)$, τ_m^2, m. It is constructed using the same rules as for the Moore EFSM $P_{C2}Y_E$. The table of LUTer2 includes 7 rows in the discussed case (Table 7.2).

This table is used to program the LUTs implementing the functions $\tau_2 = \tau_2(T)$. It is done as in the case of $P_{C2}Y_E$ Moore EFSM.

The table of EMBer includes the columns $K(a_m)$, $Y(a_m)$, τ_m^1, h, m. The column t_m^1 includes R_{CN1} bits from the codes $K(B_i)$. There is a part of this table for $P_{C2}Y_N(\Gamma_{12})$ including the first 8 rows (Table 7.3).

This table is used for programming EMBs. It is constructed in the same way as its counterpart for the case of $P_{C2}Y_E$ Moore EFSM.

Now, all tables are constructed. It gives an opportunity to obtain the NFSM logic circuit. We do not discuss this step for the Moore NFSM $P_{C2}Y_N(\Gamma_{12})$. It is possible to diminish the number of outputs of LUTer due to encoding of the inputs of NLCS $\beta_g \in C_N$. This method is based on ideas [4].

Let us discuss this approach using the GSA Γ_{13} (Fig. 7.3). It is marked by the states of Moore FSM creating the set $A = \{a_1, \ldots, a_{12}\}$.

There are the following normal LCSs in the GSA Γ_{13}: $\beta_1 = \langle a_1, a_2, a_3 \rangle$, $\beta_2 = \langle a_4, a_5, a_6 \rangle$, $\beta_3 = \langle a_7, a_8 \rangle$, $\beta_4 = \langle a_9, a_{10} \rangle$, $\beta_5 = \langle a_{11}, a_{12} \rangle$. So, there are $G2 = 5$

Table 7.3 The part of table of EMBer for Moore NFSM $P_{C2}Y_N(\Gamma_{12})$

$K(a_m)\ T_1\ldots T_4$	$Y(a_m)\ y_0\ldots y_7$	$\tau_m^1\ \tau_1$	h	m
1000	11100000	0	1	9
1001	00010000	0	2	10
1010	11100000	0	3	11
1011	00001011	1	4	12
1000	11100000	0	1	9
1001	00010000	0	2	10
1010	11100000	0	3	11
1011	00001011	1	4	12

Fig. 7.3 Initial GSA Γ_{13}

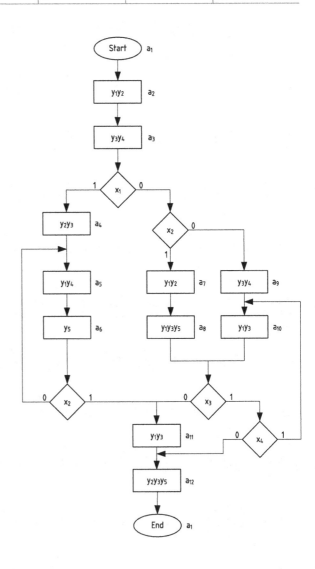

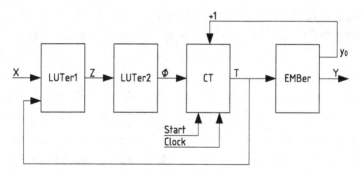

Fig. 7.4 Structural diagram of $P_{T0} Y_N$ Moore NFSM

chains in the set $C_N(\Gamma_{13})$. These chains have the following inputs: $I_1^1 = a_1$, $I_2^1 = a_4$, $I_2^2 = a_5$, $I_3^1 = a_7$, $I_4^1 = a_9$, $I_4^2 = a_{10}$, $I_5^1 = a_{11}$, $I_5^2 = a_{12}$. They form the set I_N having $|I_N| = 8$ elements.

Let us encode the inputs $I_g^k \in I_N$ by the binary codes $K(I_g^k)$ having R_{IN} bits:

$$R_{IN} = \lceil \log_2 |I_N| \rceil. \tag{7.10}$$

Let us use the variables $z_r \in Z$ for the input encoding where $|Z| = R_{IN}$. To transform the codes $K(I_g^k)$ into the codes $K(a_m)$, it is necessary to use a block of inputs transformer (BIT). The BIT implements the following system:

$$\Phi = \Phi(Z). \tag{7.11}$$

Let the symbol P_T show that there is the BIT in the structure of a Moore FSM. In the case of the base structure, four different NFSMs are possible: $P_T Y_N$, $P_{T0} Y_N$, $P_{TC1} Y_N$ and $P_{TC2} Y_N$. Design methods for these NFSMs include steps connected with design of BIT. For example, let us discuss the $P_{T0} Y_N$ Moore NFSM (Fig. 7.4).

In this structure, the LUTer1 implements the functions

$$Z = Z(T, X). \tag{7.12}$$

The LUTer2 implements the system (7.11). The LUTer2 corresponds to the BIT.

The proposed design method includes the following steps:

1. Creating the set of states A.
2. Constructing the set of NLCSs C_N.
3. Constructing the partition Π_{CN}.
4. Executing the optimal natural state assignment.
5. Executing of input encoding.
6. Constructing the table of LUTer1.

Fig. 7.5 State codes for
Moore NFSM $P_{T0}Y_N(\Gamma_{13})$

T_3T_4 \ T_1T_2	00	01	11	10
00	a_1	a_4	a_{11}	a_7
01	a_2	a_5	a_{12}	a_8
11	*	*	*	a_{10}
10	a_3	a_6	*	a_9
	B_1	B_2	B_4	B_3

Fig. 7.6 Input codes for
Moore NFSM $P_{T0}Y_N(\Gamma_{13})$

Z_3 \ Z_1Z_2	00	01	11	10
0	a_1	a_{10}	a_{12}	a_{11}
1	a_9	a_7	a_4	a_5

7. Constructing the table of LUTer2.
8. Constructing the table of EMBer.
9. Implementing the FSM logic circuit.

Let us discuss an example of design for the Moore NFSM $P_{T0}Y_N(\Gamma_{13})$. The steps 1 and 2 are already executed.

There is the partition $\Pi_{CN} = \{B_1, \ldots, B_4\}$ including classes $B_1 = \{\beta_1\}$, $B_2 = \{\beta_2\}$, $B_3 = \{\beta_3, \beta_4\}$ and $B_4 = \{\beta_5\}$. One of the variants of the optimal natural state assignment is shown in Fig. 7.5.

The following class codes can be derived from Fig. 7.5: $K(B_1) = 00^{**}$, $K(B_2) = 01^{**}$, $K(B_3) = 10^{**}$, $K(B_4) = 11^{**}$. Because the table of LUTer1 does not include the transitions for B_4, the class code $K(B_4)$ can be treated as "don't care". It allows obtaining the codes $K(B_2) = ^{*}1^{**}$ and $K(B_3) = 1^{***}$.

Using (7.10), the following value can be found: $R_{IN} = 3$. So, there is the set $Z = \{z_1, z_2, z_3\}$. Let us encode the chain inputs as it is shown in Fig. 7.6.

The table of LUTer1 is constructed on the base of the system GFT. In the discussed case, it is the system:

$$B_1 = x_1a_4 \vee \bar{x}_1x_2a_7 \vee \bar{x}_1\bar{x}_2a_9;$$
$$B_2 = x_2a_{11} \vee \bar{x}_2a_5; \tag{7.13}$$
$$B_3 = x_3x_4a_{10} \vee x_3\bar{x}_4a_{12} \vee \bar{x}_3a_{11}.$$

The table includes the columns B_i, $K(B_i)$, a_s, $K(a_s)$, X_h, Z_h, h. In the discussed case, it is Table 7.4. The state codes are taken from Fig. 7.6.

Table 7.4 Table of LUTer1 for Moore NFSM $P_{T0}Y_N(\Gamma_{13})$

B_i	$K(B_i)$	a_s	$K(a_s)$	X_h	Z_h	h
B_1	00**	a_4	111	x_1	$z_1 z_2 z_3$	1
		a_7	101	$\bar{x}_1 x_2$	$z_1 z_3$	2
		a_9	100	$\bar{x}_1 \bar{x}_2$	z_1	3
B_2	*1**	a_{11}	010	x_2	z_2	4
		a_5	110	$\bar{x}_2$	$z_1 z_2$	5
B_3	1***	a_{10}	001	$x_3 z_4$	z_3	6
		a_{12}	011	$x_3 \bar{x}_4$	$z_2 z_4$	7
		a_{11}	010	$\bar{x}_3$	z_2	8

Table 7.5 Table of LUTer2 for Moore NFSM $P_{T0}Y_N(\Gamma_{13})$

a_m	$C(a_m) z_1 z_2 z_3$	$K(B_i) T_1 \ldots T_4$	Φ_h	m
a_1	000	0000	–	1
a_4	111	0100	D_2	4
a_5	110	0101	$D_2 D_4$	5
a_7	101	1000	D_1	7
a_9	100	0110	$D_1 D_3$	9
a_{10}	001	1011	$D_1 D_3 D_4$	10
a_{11}	010	1100	$D_1 D_2$	11
a_{12}	011	1101	$D_1 D_2 D_4$	10

The table of LUTer1 is used for deriving the system (7.12). It is the following system in the discussed example:

$$z_1 = \bar{T}_1 \bar{T}_2 \vee T_2 \bar{x}_2;$$
$$z_2 = \bar{T}_1 \bar{T}_2 x_1 \vee T_2 \vee T_1 \bar{x}_4 \vee T_1 \bar{x}_3; \tag{7.14}$$
$$z_3 = \bar{T}_1 \bar{T}_2 x_1 \vee \bar{T}_1 \bar{T}_2 x_2 \vee T_1 x_3.$$

Each of Eqs. (7.14) can be implemented using LUTs with $S_L = 5$.

The table of LUTer2 gives functions (7.11). It includes the columns a_m, $C(a_m)$, $K(a_m)$, Φ_m, m. In the discussed case, it is Table 7.5.

The state codes $C(a_m)$ are taken from Fig. 7.6, whereas the state codes $K(a_m)$ from Fig. 7.5. If the following relation takes place

$$S_L \geq R_{IN}, \tag{7.15}$$

then only R_{IN} of LUTs is required to implement the logic circuit of LUTer2.

The table of EMBer is the same as for PY_E Moore EFSM. For the discussed example, a part of this table is represented by Table 7.6.

Table 7.6 The part of table of EMBer for Moore NFSM $P_{T0}Y_N(\Gamma_{13})$

$K(a_m)$ $T_1 \ldots T_4$	$Y(a_m)$ $y_0 \ldots y_5$	h	m
1000	100000	1	1
0001	111000	2	2
0010	000110	3	3
0011	000000	4	*
0100	101100	5	4

Fig. 7.7 Structural diagram of FPGA-based PY_{NC} Moore NFSM

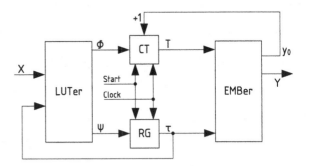

Let us use LUTs having $S_L = 5$ and EMBs having the configuration 16×8. In this case, each equation of (7.14) is implemented using only a single LUT. Because the (7.15) takes place, each Eq. (7.11) is implemented using only a single LUT. So, there are seven LUTs and one EMB in the logic circuit of Moore NFSM $P_{T0}Y_N(\Gamma_{13})$. Of course, at least four LUTs are used to implement the circuit of CT.

This approach can be used in ELCS-based Moore FSMs. It results in models $P_T Y_E$, $P_{T0} Y_E$, $P_{TC1} Y_E$, $P_{TC2} Y_E$. We do not consider them in our book.

7.2   Optimization of NFSMs with Code Sharing

Basing on Fig. 3.17, it can be obtained the FPGA-based structural diagram of PY_{NC} Moore NFSM (Fig. 7.7).

In this model, the LUTer implements functions (3.57) and (3.58), the EMBer generates functions (3.59) and (3.60). The counter CT contains codes $C(a_m)$ of states $a_m \in A$. The register RG contains codes $K(\beta_g)$ of chains $\beta_g \in C_N$. The state codes $K(a_m)$ are represented using the principle of code sharing (3.56).

The following approach is proposed for synthesis of FPGA-based PY_{NC} Moore FSM:

1. Creating the set of states A for a given GSA Γ.
2. Constructing the set of NLCSs $C_N = \{\beta_1, \ldots, \beta_{G2}\}$.
3. Executing encoding of NLCS.
4. Executing natural encoding (3.55).

Table 7.7 Natural LCSs and their classes for GSA Γ_{14}

B_i	B_1	B_2			B_3			B_4	$C(a_m)$
β_g	β_1	β_2	β_3	β_4	β_5	β_6	β_7	β_8	
a_m	a_1	a_3	a_6	a_8	a_{10}	a_{11}	a_{13}	a_{15}	00
	a_2	a_4	a_7	a_9	–	a_{12}	a_{14}	a_{16}	01
	–	a_5	–	–	–	–	–	a_{17}	10
	–	–	–	–	–	–	–	–	11
$K(\beta_g)$	000	001	010	011	100	101	110	111	–
$K(B_i)$	00	01			10			11	–

5. Constructing the table of LUTer.
6. Constructing the table of EMBer.
7. Implementing the FSM logic circuit.

Let us discuss an example of synthesis for the More NFSM $PY_{NC}(\Gamma_{14})$. The marked GSA Γ_{14} is shown in Fig. 7.8.

The following sets and their characteristics can be found from GSA Γ_{14}: $X = \{x_1, \ldots, x_4\}$, $L = 4$, $Y = \{y_1, \ldots, y_7\}$, $N = 7$, $A = \{a_1, \ldots, a_{17}\}$, $M = 17$. Hence, there is $R = 5$.

Let us apply the procedure P_4 to this GSA. It gives the set $C_N = \{\beta_1, \ldots, \beta_8\}$ where $\beta_1 = \langle a_1, a_2 \rangle$, $\beta_2 = \langle a_3, a_4, a_5 \rangle$, $\beta_3 = \langle a_6, a_7 \rangle$, $\beta_4 = \langle a_8, a_9 \rangle$, $\beta_5 = \langle a_{10} \rangle$, $\beta_6 = \langle a_{11}, a_{12} \rangle$, $\beta_7 = \langle a_{13}, a_{14} \rangle$ and $\beta_8 = \langle a_{15}, a_{16}, a_{17} \rangle$. So, there is $G2 = 3$. It gives the set $\tau = \{\tau_1, \tau_2, \tau_3\}$. Analysis of chains shows that there is $M_{G2} = 3$. It gives the value $R_{C2} = 2$ and the set $T = \{T_1, T_2\}$. Also, there are set $\Psi = \{D_1, D_2, D_3\}$ and $\Phi = \{D_4, D_5\}$.

Let us encode the chains $\beta_g \in C_N$ in the trivial way: $K(\beta_1) = 000$, $K(\beta_2) = 001, \ldots, K(\beta_8) = 111$. Let us encode the states $a_m \in A(\beta_g)$ using (3.55). The resulting codes $K(a_m)$ can be obtained from Table 7.7.

Using Table 7.7, the following state codes, for example, can be found: $K(a_1) = 00000$, $K(a_2) = 00001$, $K(a_{17}) = 11110$.

The table of LUTer includes the columns β_g, $K(\beta_g)$, a_s, $K(a_s)$, X_h, Ψ_h, Φ_h, h. The sense of them is clear from previous discussion. The table is constructed on the base of the system of formulae of transitions. In the discussed case, it is the following system:

$$a_2 \rightarrow x_1 x_3 a_3 \vee x_1 \bar{x}_3 a_4 \vee \bar{x}_1 x_2 a_6 \vee \bar{x}_1 \bar{x}_2 a_8;$$

$$a_5, a_7, a_9 \rightarrow x_3 a_{10} \vee \bar{x}_3 x_4 a_{11} \vee \bar{x}_3 \bar{x}_4 x_1 a_{13} \vee \bar{x}_1 \bar{x}_3 \bar{x}_4 a_{15};$$

$$a_{10}, a_{12}, a_{14} \rightarrow a_1;$$

$$a_{17} \rightarrow x_3 a_{14} \vee \bar{x}_3 a_9. \tag{7.16}$$

In the discussed case, the table of LUTer includes 18 rows. Let us point out that there are no transitions for states a_{10}, a_{12} and a_{14} in the table. These transitions are

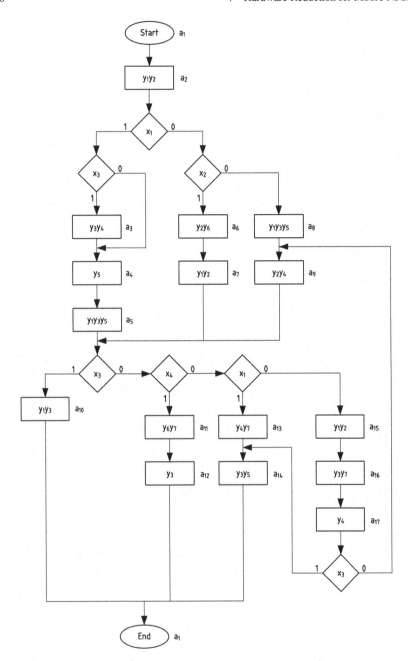

Fig. 7.8 Initial marked GSA Γ'_{14}

Table 7.8 The part of table of LUTer for Moore NFSM $PY_{NC}(\Gamma_{14})$

β_g	$K(\beta_g)$	a_s	$K(a_s)$	X_h	Ψ_h	Φ_h	h
β_1	000	a_3	00100	$x_1 x_3$	D_3	–	1
		a_4	00101	$x_1 \bar{x}_3$	D_3	D_5	2
		a_6	01000	$\bar{x}_1 x_2$	D_2	–	3
		a_8	01100	$\bar{x}_1 \bar{x}_2$	$D_2 D_3$	–	4
β_2	001	a_{10}	10000	x_3	D_1	–	5
		a_{11}	10100	$\bar{x}_3 x_4$	$D_1 D_3$	–	6
		a_{13}	11000	$x_1 \bar{x}_3 \bar{x}_4$	$D_1 D_2$	–	7
		a_{15}	11100	$\bar{x}_1 \bar{x}_3 \bar{x}_4$	$D_1 D_2 D_3$	–	8

Table 7.9 The part of table of EMBer for Moore NFSM $P_0 Y_{NC}(\Gamma_{14})$

$K(a_m) \, \tau_1 \tau_2 \tau_3 T_1 T_2$	$Y(a_m) \, y_0 \ldots y_7$	h	m
000 00	1000 0000	1	1
000 01	0110 0000	2	2
000 10	0000 0000	3	*
000 11	0000 0000	4	*
001 00	1001 1000	5	3
001 01	1000 0100	6	4
001 10	0101 0100	7	5
001 11	0000 0000	8	*

executed automatically. First 8 rows are shown in Table 7.8 for the table of LUTer of Moore NFSM $PY_{NC}(\Gamma_{14})$.

The table of LUTer gives functions (3.57) and (3.58). For example, the following equations can be derived from Table 7.8 (after minimization):

$$
\begin{aligned}
D_1 &= \bar{\tau}_1 \bar{\tau}_2 \tau_3; \\
D_2 &= \bar{\tau}_1 \bar{\tau}_2 \bar{\tau}_3 \bar{x}_1 \vee \bar{\tau}_1 \bar{\tau}_2 \tau_3 \bar{x}_3 \bar{x}_4; \\
D_3 &= \bar{\tau}_1 \bar{\tau}_2 \tau_3 x_1 \vee \bar{\tau}_1 \bar{\tau}_2 \tau_3 \bar{x}_2 \vee \bar{\tau}_1 \bar{\tau}_2 \tau_3 (\bar{x}_3 x_4 \vee \bar{x}_1 \bar{x}_3 \bar{x}_4); \\
D_4 &= 0; \\
D_5 &= \bar{\tau}_1 \bar{\tau}_2 \bar{\tau}_3 x_1 \bar{x}_3.
\end{aligned}
\tag{7.17}
$$

The table of EMBer includes the columns $K(a_m)$, $Y(a_m)$, h, m. It is constructed in the same way as for its counterpart for PY_{EC} Moore EFSM. The first 8 lines are shown in Table 7.9 for the given example. Let us point out that there are 32 rows in the table of EMBer for our example.

T_1T_2 \ $t_1t_2t_3$	000	001	011	010	110	111	101	100
00	a_1	a_{15}	a_{11}	a_{13}	a_3	a_6	a_8	a_{10}
01	a_2	a_{16}	a_{12}	a_{14}	a_4	a_7	a_9	✳
11	✳	✳	✳	✳	✳	✳	✳	✳
10	✳	a_{17}	✳	✳	a_5	✳	✳	✳

Fig. 7.9 Optimal natural state codes for Moore NFSM $P_0 Y_{NC}(\Gamma_{14})$

Basing on Chap. 6, two methods can be used for optimization of logic circuit of PY_{NC} Moore FSM:

1. The optimal natural state assignment. It leads to $P_0 Y_{NC}$ Moore NFSM having the same structure as the one shown in Fig. 7.6.
2. The transformation of chain codes into class codes. There are two modification of this FSM. They depends on the following condition

$$t_F > N + 1 + R_{CN}. \tag{7.18}$$

The value of R_{CN} determines the number of bits in the class codes $K(B_i)$.

There are the optimal natural state codes for the Moore NFSM $P_0 Y_{NC}(\Gamma_{14})$ shown in Fig. 7.9.

The partition Π_{CN} is taken from Table 7.7. It includes the classes $B_1 = \{\beta_1\}$, $B_2 = \{\beta_2, \beta_3, \beta_4\}$, $B_3 = \{\beta_5, \beta_6, \beta_7\}$, $B_4 = \{\beta_8\}$. We can use the codes of states a_{10}–a_{14} for optimizing the class codes. Using this fact, the following class codes can be found from Fig. 7.9: $K(B_1) = 0*0**$, $K(B_2) = 1****$ and $K(B_4) = 0*1**$.

The system of GFT for $P_0 Y_{NC}(\Gamma_{14})$ is similar to the system (7.16). But the states $a_m = O_g$ should be replaced by corresponding classes:

$$
\begin{aligned}
B_1 &\rightarrow x_1 x_3 a_3 \vee x_1 \bar{x}_3 a_4 \vee \bar{x}_1 x_2 a_6 \vee \bar{x}_1 \bar{x}_2 a_8; \\
B_2 &\rightarrow x_3 a_{10} \vee \bar{x}_3 x_4 a_{11} \vee x_1 \bar{x}_3 x_4 a_{13} \vee \bar{x}_3 \bar{x}_4 a_{15}; \\
B_4 &\rightarrow x_3 a_{14} \vee \bar{x}_3 a_9.
\end{aligned}
\tag{7.19}
$$

The system (7.19) determines Table 7.10. It includes only 10 rows.

As in the previous case, functions (3.57) and (3.58) can be derived from this table. For example, the following functions can be derived from Table 7.10 (after minimization):

$$
\begin{aligned}
D_1 &= \bar{\tau}_1 \bar{\tau}_3 \vee \tau_1 x_3 \vee \tau_1 x_4 \vee \bar{\tau}_1 \tau_3 \bar{x}_3; \\
D_5 &= \bar{\tau}_1 \bar{\tau}_3 x_1 \bar{x}_3 \vee \bar{\tau}_1 \tau_3.
\end{aligned}
\tag{7.20}
$$

Table 7.10 Table of LUTer for Moore NFSM $P_0 Y_{NC}(\Gamma_{14})$

B_i	$K(B_i)$	a_s	$K(a_s)$	X_h	Ψ_h	Φ_h	h
B_1	0*0**	a_3	11000	$x_1 x_3$	$D_1 D_2$	–	1
		a_4	11001	$x_1 \bar{x}_3$	$D_1 D_2$	D_5	2
		a_6	11100	$\bar{x}_1 x_2$	$D_1 D_2 D_3$	–	3
		a_8	10100	$\bar{x}_1 \bar{x}_2$	$D_1 D_3$	–	4
B_2	1****	a_{10}	10000	x_3	D_1	–	5
		a_{11}	01100	$\bar{x}_3 x_4$	$D_1 D_2$	–	6
		a_{13}	01000	$x_1 \bar{x}_3 \bar{x}_4$	D_2	–	7
		a_{15}	00100	$\bar{x}_1 \bar{x}_3 \bar{x}_4$	D_3	–	8
B_4	0*1**	a_{14}	01001	x_3	D_2	D_5	9
		a_9	10101	$\bar{x}_3$	$D_1 D_3$	D_5	10

Fig. 7.10 Structural diagram of FPGA-based $P_{C1} Y_{NC}$ Moore NFSM

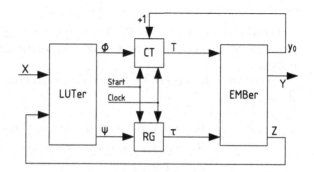

If the condition (7.18) takes place, then the model of $P_{C1} Y_{NC}$ Moore NFSM can be used (Fig. 7.10).

In this model, the LUTer implements the systems (6.30) and

$$\Phi = \Phi(Z, X). \tag{7.21}$$

The EMBer implements the functions (3.59), (3.60) and (3.62). The proposed design method for $P_{C1} Y_{NC}$ Moore NFSM includes the following steps:

1. Creating the set of states A.
2. Constructing the set of NLCSs C_N.
3. Executing chain encoding and natural state encoding (3.55).
4. Constructing the set Π_{NC}.
5. Executing class encoding.
6. Constructing the table of LUTer.
7. Constructing the table of EMBer.
8. Implementing the FSM logic circuit.

Let us discuss an example of design fot the Moore NFSM $P_{C1} Y_{NC}(\Gamma_{14})$. The first five steps are already executed. Their outcomes is represented by Table 7.7.

Table 7.11 Table of LUTer for Moore NFSM $P_{C1}Y_{NC}(\Gamma_{14})$

B_i	$K(B_i)$	a_s	$K(a_s)$	X_h	Ψ_h	Φ_h	h
B_1	00	a_3	00100	$x_1 x_3$	D_3	–	1
		a_4	00101	$x_1 \bar{x}_3$	D_3	D_5	2
		a_6	01000	$\bar{x}_1 x_2$	D_2	–	3
		a_8	01100	$\bar{x}_1 \bar{x}_2$	$D_2 D_3$	–	4
B_2	01	a_{10}	10000	x_3	D_1	–	5
		a_{11}	10100	$\bar{x}_3 x_4$	$D_1 D_3$	–	6
		a_{13}	11000	$x_1 \bar{x}_3 \bar{x}_4$	$D_1 D_2$	–	7
		a_{15}	11100	$\bar{x}_1 \bar{x}_3 \bar{x}_4$	$D_1 D_2 D_3$	–	8
B_4	11	a_{14}	10001	x_3	$D_1 D_2$	D_5	9
		a_9	01101	$\bar{x}_3$	$D_2 D_3$	D_5	10

Let us point out that there are $R_{CN} = 2$ and $Z = \{z_1, z_2\}$. The code 10 can be treated as "don't care". It can be done due to execution of transitions $\langle a_m, a_1 \rangle$ automatically.

To construct the table of LUTer, let us construct the system of GFT for classes B_1, B_2, and B_4:

$$
\begin{aligned}
B_1 &\to x_1 x_3 a_3 \vee x_1 \bar{x}_3 a_4 \vee \bar{x}_1 x_2 a_6 \vee \bar{x}_1 \bar{x}_2 a_8; \\
B_2 &\to x_3 a_{10} \vee \bar{x}_3 x_4 a_{11} \vee \bar{x}_3 \bar{x}_4 x_1 a_{13} \vee \bar{x}_1 \bar{x}_3 \bar{x}_4 a_{15}; \\
B_4 &\to x_3 a_{14} \vee \bar{x}_3 a_9.
\end{aligned}
\tag{7.22}
$$

The system (7.21) determines Table 7.11.

Let us point out that the chain codes are taken from Table 7.7. The table of LUTer is used for deriving the systems (6.30) and (7.21). For example, it is possible to derive the following functions from Table 7.11:

$$
\begin{aligned}
D_1 &= \bar{z}_1 z_2 \vee z_1 z_2 x_3; \\
D_2 &= \bar{z}_1 \bar{z}_2 \bar{x}_1 \vee z_1 z_2 \bar{x}_3 \bar{x}_4 \vee z_1 z_2; \\
D_3 &= \bar{z}_1 \bar{z}_2 x_1 \vee z_1 \bar{z}_2 \bar{x}_2 \vee z_1 z_2 \bar{x}_3 x_4 \vee \bar{z}_1 x_2 x_1 \bar{x}_3 \bar{x}_4 \vee z_1 z_2 \bar{x}_3; \\
D_4 &= 0; \\
D_5 &= \bar{z}_1 \bar{z}_2 x_1 \bar{x}_3 \vee z_1 z_2.
\end{aligned}
\tag{7.23}
$$

Each equation of (7.23) can be implemented using a single LUT having 6 inputs.

To implement the circuit of EMBer, it is necessary to use an EMB having the configuration 32×10 (bits). The table of EMBer is constructed in a trivial way. The logic circuit of NFSM $P_{C1}Y_{NC}(\Gamma_{14})$ is shown in Fig. 7.11.

If the condition (7.18) is violated, then the model of $P_{C2}Y_{NC}$ Moore NFSM can be used (Fig. 7.12).

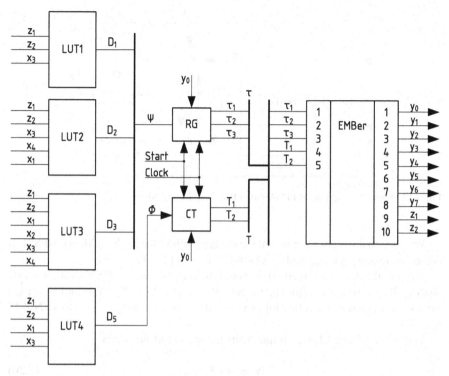

Fig. 7.11 Logic circuit of Moore NFSM $P_{C1}Y_{NC}(\Gamma_{14})$

Fig. 7.12 Structural diagram
of $P_{C2}Y_{NC}$ Moore NFSM

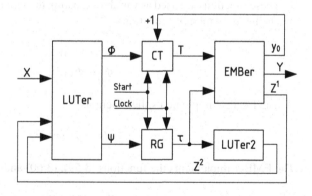

It can be used if the following conditions take places:

$$t_F \geq N + 1; \tag{7.24}$$

$$t_f < N + 1 + R_{CN} \tag{7.25}$$

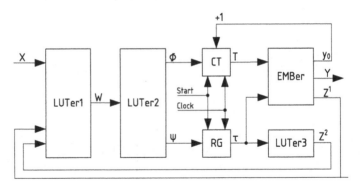

Fig. 7.13 Structural diagram of FPGA-based $P_{TC2}Y_{NC}$ Moore NFSM

We hope that a reader can invent the design method for $P_{C2}Y_{NC}$ Moore NFSM. It can be done using the methods for both $P_{C2}Y_{EC}$ and $P_{C1}Y_{EC}$ FSMs.

It is possible to use the method of input transformations for NFSMs with the code sharing. It results in the following models: $P_T Y_{NC}$, $P_{T0}Y_{NC}$, $P_{TC1}Y_{NC}$ and $P_{TC2}Y_{NC}$. For example, there is the following structure diagram for the $P_{TC2}Y_{NC}$ Moore NFSM (Fig. 7.13).

In this model, the LUTer1 implements the system of functions

$$W = W(Z, X). \tag{7.26}$$

These functions are used as variables creating the input codes. The LUTer2 implements the input memory functions

$$\Psi = \Psi(W); \tag{7.27}$$
$$\Phi = \Phi(W). \tag{7.28}$$

The LUTer3 implements the functions

$$Z^2 = Z^2(\tau). \tag{7.29}$$

The EMBer implements the functions (3.59), (3.60) and

$$Z^1 = Z^1(\tau, T). \tag{7.30}$$

The variables $z_r =\in Z^1 \cup Z^2$ are used for representing the class codes.

There are the following steps in the proposed design method for $P_{TC2}Y_{NC}$ Moore NFSM:

1. Creating the set of states A.
2. Constructing the set of chains C_N.
3. Executions of state and chain encoding.

Fig. 7.14 Codes of inputs
for Moore NFSM
$P_{TC2}Y_{NC}(\Gamma_{14})$

W_3W_4 \ W_1W_2	00	01	11	10
00	a_1	a_{10}	*	a_3
01	a_9	a_{13}	*	a_6
11	a_{14}	a_{15}	*	a_8
10	*	a_{11}	*	a_4

4. Constructing the set of Π_{NC}.
5. Executing the class encoding.
6. Executing chain inputs encoding.
7. Constructing the table of LUTer1.
8. Constructing the table of LUTer2.
9. Constructing the table of LUTer3.
10. Constructing the table of EMBer.
11. Implementing the FSM logic circuit.

Let us discuss an example of design for the Moore NFSM $P_{TC2}Y_{NC}(\Gamma_{14})$. The first five steps are already executed. There are the corresponding sets and codes in Table 7.7.

It is possible to find the set of inputs for NLCSs of GSA Γ_{14}: $I_N = \{a_1, a_2, a_4, a_6, a_8, a_9, a_{10}, a_{11}, a_{13}, a_{14}, a_{15}\}$. It is necessary $R_{IN} = 4$ variables to encode the inputs. Let us encode the inputs as it is shown in Fig. 7.14. The following approach is used for this step: all states from the same GFT should be included in the same generalized interval of R_{IN}-dimensional Boolean space. It allows minimization for Boolean equations (7.26).

The table of LUTer1 (Table 7.12) is constructed using the system of GFT (7.22). It contains the state codes from Fig. 7.14.

Now, the following equations can be derived from Table 7.12:

$$
\begin{aligned}
w_1 &= \bar{z}_1 \bar{z}_2; \\
w_2 &= \bar{z}_1 z_2; \\
w_3 &= \bar{z}_1 z_2 x_1 x_3 \vee \bar{z}_1 \bar{z}_2 \bar{x}_1 \bar{x}_2 \vee \bar{z}_1 z_2 \bar{x}_3 x_4 \vee \bar{z}_1 z_2 \bar{x}_1 \bar{x}_3 \bar{x}_4 \vee z_1 z_2 \bar{x}_3; \\
w_4 &= \bar{z}_1 \bar{z}_2 \bar{x}_1 \vee \bar{z}_1 z_2 \bar{x}_3 \bar{x}_4 \vee z_1 z_2.
\end{aligned}
\tag{7.31}
$$

The table of LUTer2 includes the columns: a_m, $K(a_m)$, $K(\beta_g)$, $C(a_m)$, Ψ_h, Φ_h, h. It includes the codes of inputs in the column $K(a_m)$. It is used for deriving the functions (7.27) and (7.28). In the discussed case it is Table 7.13.

Let us point out that there is no row for state a_1 in Table 7.13. The pulse *Start* is used for loading the zero codes into both RG and CT. Table 7.13 is used for deriving

Table 7.12 Table of LUTer1 for Moore NFSM $P_{TC2}Y_{NC}(\Gamma_{14})$

B_i	$K(B_i)$	a_s	$K(a_s)$	X_h	W_h	h
B_1	00	a_3	1000	$x_1 x_3$	w_1	1
		a_4	1010	$x_1 \bar{x}_3$	$w_1 w_3$	2
		a_6	1001	$\bar{x}_1 \bar{x}_2$	$w_1 w_4$	3
		a_8	1011	$\bar{x}_1 \bar{x}_2$	$w_1 w_3 w_4$	4
B_2	01	a_{10}	0100	x_3	w_2	5
		a_{11}	0110	$\bar{x}_3 x_4$	$w_2 w_3$	6
		a_{13}	0101	$x_1 \bar{x}_3 \bar{x}_4$	$w_2 w_4$	7
		a_{15}	0111	$\bar{x}_1 \bar{x}_3 \bar{x}_4$	$w_2 w_3 w_4$	8
B_4	11	a_{14}	0001	x_3	w_4	9
		a_9	0011	$\bar{x}_3$	$w_3 w_4$	10

Table 7.13 Table of LUTer2 for Moore NFSM $P_{TC2}Y_{NC}(\Gamma_{14})$

a_m	$K(a_m)$	$K(\beta_g)$	$C(a_m)$	Ψ_h	Φ_h	h
a_3	1000	001	00	D_3	–	1
a_4	1010	001	01	D_3	D_5	2
a_6	1001	010	00	D_2	–	3
a_8	1011	011	00	$D_2 D_3$	–	4
a_9	0001	011	01	$D_2 D_3$	D_5	5
a_{10}	0100	100	00	D_1	–	6
a_{11}	0110	101	00	$D_1 D_3$	–	7
a_{13}	0101	110	00	$D_1 D_2$	–	8
a_{14}	0011	110	01	$D_1 D_2$	D_5	9
a_{15}	0111	111	00	$D_1 D_2 D_3$	–	10

the functions (7.27) and (7.28). For example, the following equations can be derived from Table 7.13:

$$D_1 = \bar{w}_1 w_2 \vee \bar{w}_1 w_3;$$
$$D_5 = w_1 w_3 \bar{w}_4 \vee \bar{w}_1 \bar{w}_2 w_3. \tag{7.32}$$

These equations are obtained using both Table 7.13 and the Karnaugh map (Fig. 7.14).

Let the EMB in use have the configuration 32×9 bits. Because of $I_N = 4$, we have $R_{CN} = 2$. So, both conditions (7.24) and (7.25) take place. Let us divide the set Z by the following subsets: $Z^1 = \{z_2\}$ and $Z^2 = \{z_1\}$. Analysis of class codes shows that $z_1 = 1$ only for chain $\beta_8 \in B_4$. So there is no need in the table of LUTer3. We can obtain the following equation $z_1 = \tau_1 \tau_2 \tau_3$ from the relation $\beta_8 \in B_4$.

The table of EMBer is constructed as in previous cases. We do not discuss this step in that book.

Table 7.14 Models of NLCS-based Moore FSMs

No.	Type	BIMF	BTC	BIT	BMO
1.	PY_N	$\Phi = \Phi(T, X)$ [1]	–	–	$y_0 = y_0(T)$ [2]
2.	P_0Y_N				$Y = Y(T)$ [3]
3.	$P_{C1}Y_N$	$\Phi = \Phi(\tau, X)$ [4]	–	–	[2][3] $\tau = \tau(T)$ [5]
4.	$P_{C2}Y_N$		$\tau^2 = \tau^2(T)$ [6]	–	[2][3] $\tau^1 = \tau^1(T)$ [7]
5.	P_TY_N	$Z = Z(T, X)$ [8]	–	$\Phi = \Phi(Z)$ [9]	[2][3]
6.	$P_{T0}Y_N$				
7.	$P_{TC1}Y_N$	$Z = Z(\tau, X)$ [10]	–		[2][3][5]
8.	$P_{TC2}Y_N$		[6]		[2][3][7]
9.	PY_{NC}	$\Psi = \Psi(T, X)$ [11]	–	–	$y_0 = y_0(\tau, T)$ [13]
10.	P_0Y_{NC}	$\Phi = \Phi(T, X)$ [12]			$Y = Y(\tau, T)$ [14]
11.	$P_{C1}Y_{NC}$	$\Psi = \Psi(Z, X)$ [15]	–	–	[13][14] $Z = Z(\tau, T)$ [17]
12.	$P_{C2}Y_{NC}$	$\Phi = \Phi(Z, X)$ [16]	$Z^2 = Z^2(\tau)$ [18]		[13][14] $Z^1 = Z^1(\tau, T)$ [19]
13.	P_TY_{NC}	$W = W(\tau, X)$ [20]	–	$\Phi = \Phi(W)$ [21]	[13][14]
14.	$P_{T0}Y_{NC}$			$\Psi = \Psi(W)$ [22]	
15.	$P_{TC1}Y_{NC}$	$W = W(Z, X)$ [23]	–	–	[13][14][17]
16.	$P_{TC2}Y_{NC}$		[18]		[13][14][19]

There are 16 different models of NFSMs discussed in Sects. 7.1 and 7.2. They are represented by Table 7.14. All these models target FPGA as a basis for implementing logic circuits of NFSMs.

The further hardware reduction can be obtained due to replacement of logical conditions [2]. These methods can be used for FSMs based on any kind of LCSs. Let us discuss these methods for Moore NFSMs.

7.3 Replacement of Logical Conditions for NLCS-based Moore FSMs

As it is mentioned in Chap. 2, the replacement of logical conditions is reduced to finding some set of additional variables $P = \{p_1, \ldots, p_G\}$. The value of G is determined as $\max(L_1, \ldots, L_M)$. The symbol L_m stands for the number of logical conditions in the state $X(a_m) \subset X$. The logical conditions $x_e \in X(a_m)$ determine transitions from the state $a_m \in A$. If the method of RLC is used, then the symbol "M" appears in the corresponding formula of FSM.

Let us start from MPY_N Moore NFSM. There is a structural diagram of FPGA-based MPY_N Moore NFSM shown in Fig. 7.15.

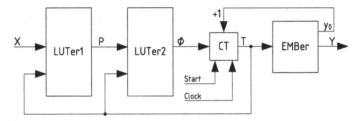

Fig. 7.15 Structural diagram of FPGA-based MPY$_N$ Moore NFSM

Table 7.15 Natural LCSs and their characteristics for GSA Γ_{13}

B_i	B_1	B_2	B_3		B_4	$C(a_m)$
β_g	β_1	β_2	β_3	β_4	β_5	
a_m	a_1	a_4	a_7	a_9	a_{11}	00
	a_2	a_5	a_8	a_{10}	a_{12}	01
	a_3	a_6	–	–	–	10
$K(\beta_g)$	000	001	010	011	100	–
$K(B_i)$	00	01	10		11	–

In this model, the LUTer1 implements the system (2.19), whereas, the LUTer2 generates the input memory functions (2.20). The EMBer implements functions (2.16) and (3.29). The proposed design methods includes the following steps:

1. Constructing the set of states A.
2. Constructing the set of chains C_N.
3. Executions of the natural state assignment.
4. Finding the set of additional variables P.
5. Constructing the table of LUTer1.
6. Constructing the table of LUTer2.
7. Constructing the table of EMBer.
8. Implementing the FSM logic circuit.

Let us discuss an example of design for the Moore NFSM MPY$_N(\Gamma_{13})$. The GSA Γ_{13} is shown in Fig. 7.3. There are chains, classes and their codes in Table 7.15.

Let us form the set $X(a_m)$ for states $a_m = O_g(g = 1, \ldots, G2)$. There are the following states $X(a_3) = \{x_1, x_2\}$, $X(a_6) = \{x_2\}$, $X(a_8) = X(A_{10}) = \{x_3, x_4\}$, $X(a_{12}) = \emptyset$. Obviously, there is $G = 2$. It gives the set $P = \{p_1, p_2\}$.

Let us form the table showing the replacement of logical conditions. It is Table 7.16 for the given example.

Let us encode the states $a_m \in A$ in the trivial way: $K(a_1) = 0000$, $K(a_2) = 0001, \ldots, K(a_{12}) = 1011$. It corresponds to the requirement (7.1). The table of LUTer1 has the following columns: a_m, $K(a_m)$, $X(p_1), \ldots, X(P_G)$, m. It is constructed on the base of table of RLC. In the discussed case, it is Table 7.16. Let us point out that there are the sets $X(p_1) = \{x_1, x_3\}$, $X(p_2) = \{x_2, x_4\}$.

Table 7.16 Table of RLC for Moore NFSM $MPY_N(\Gamma_{13})$

$P_g \backslash a_m$	a_3	a_6	a_8	a_{10}	a_{12}
P_1	x_1	–	x_3	x_3	–
P_2	x_2	x_2	x_4	x_4	–

Table 7.17 Table of LUTer1 for Moore NFSM $MPY_{CN}(\Gamma_{13})$

a_m	$K(a_m)$	$X(p_1)$	$X(p_2)$	m
a_8	0111	x_3	x_4	8
a_{10}	1001	x_3	x_4	10
a_3	0010	x_1	x_2	3
a_6	0101	–	x_2	6

Fig. 7.16 Refined state codes for Moore NFSM $MPY_{CN}(\Gamma_{13})$

T_3T_4 \ T_1T_2	00	01	11	10
00	a_1	a_4	a_{11}	a_7
01	a_2	a_5	a_{12}	a_8
11	*	*	*	a_{10}
10	a_3	a_6	*	a_9

This table is used for deriving Eqs. (2.19). In the discussed case, there is the system:

$$p_1 = \bar{T}_1\bar{T}_2T_3\bar{T}_4x_1 \vee \bar{T}_1T_2T_3T_4x_3 \vee T_1\bar{T}_2\bar{T}_3T_4x_3;$$
$$p_2 = \bar{T}_1\bar{T}_2T_3\bar{T}_4x_2 \vee \bar{T}_1T_2\bar{T}_3T_4x_2 \vee \ldots \vee T_1\bar{T}_2\bar{T}_3T_4x_4. \tag{7.33}$$

Each equation of (7.33) can be implemented using a single LUT with $S_L = 6$. These equations can be minimized. If $a_m \neq O_g$, then the code $K(a_m)$ can be used for minimizing. The same is true for the state a_m such that there is unconditional transition $\langle a_m, a_1 \rangle$ (Table 7.17).

Let us change the state codes for $MPY_{CN}(\Gamma_{13})$. Let new codes be oriented on optimization of the system (2.19). Let us name such an approach as a refined state assignment. One of the possible variants is shown in Fig. 7.16.

Using these state codes and "don't care" state codes, the following system can be obtained:

$$p_1 = \bar{T}_1x_1 \vee T_1x_3;$$
$$p_2 = \bar{T}_1x_2 \vee T_1x_4. \tag{7.34}$$

Table 7.18 Table of LUTer2 for Moore NFSM $MPY_{CN}(\Gamma_{13})$

a_m	$K(a_m)$	a_s	$K(a_s)$	P_h	Φ_h	h
a_3	0010	a_4	0100	p_1	D_2	1
		a_7	1000	$\bar{p}_1 p_2$	D_1	2
		a_9	1010	$\bar{p}_1 \bar{p}_2$	$D_1 D_3$	3
a_6	0110	a_{11}	1100	p_2	$D_1 D_2$	4
		a_5	0101	$\bar{p}_2$	$D_2 D_4$	5
a_8	1001	a_{12}	1101	$p_1 \bar{p}_2$	$D_1 D_2 D_4$	6
		a_{10}	1011	$p_1 p_2$	$D_1 D_3 D_4$	7
		a_{11}	1100	$\bar{p}_1$	$D_1 D_2$	8
a_{10}	1011	a_{12}	1101	$p_1 \bar{p}_2$	$D_1 D_2 D_4$	9
		a_{10}	1011	$p_1 p_2$	$D_1 D_3 D_4$	10
		a_{11}	1100	$\bar{p}_1$	$D_1 D_2$	11

Each equation of (7.34) can be implemented with a single LUT having $S_L = 3$. Let us point out that the codes from Fig. 7.16 satisfy to (7.1).

The table of LUTer2 includes the following columns: a_m, $K(a_m)$, a_s, $K(a_s)$, P_h, Φ_h, h. The column P_h includes conjunctions of variables $p_g \in P$ corresponding to conjunctions X_h for $PY_N(\Gamma_{13})$. In the discussed case, it is Table 7.18.

The state codes are taken from Fig. 7.16. The equations (2.30) can be derived from table of LUTer2. For example, the following minimized equations can be derived from Table 7.18:

$$D_1 = \bar{T}_1 \bar{T}_2 T_3 \bar{p}_1 \vee T_2 T_3 p_2 \vee T_1 \bar{T}_2 T_4;$$
$$D_4 = T_2 T_3 \bar{p}_2 \vee T_1 \bar{T}_2 p_1. \tag{7.35}$$

Let us point out that each equation $D_r \in \Phi$ can contain up to $R + G$ different variables. In the case of PY_N Moore NFSM, it contains up to $R + L$ variables. If $L \gg G$, then the circuit of LUTer2 is quite simpler that the circuit of LUTer for equivalent PY_N Moore NFSM.

Tables of EMBer are the same for MPY_N and PY_N Moore NFSMs. Because of it, we do not discuss the step 7 in our book.

The method of RLC can be used together with the encoding of inputs of NLCSs. Using both methods the following structural diagram (Fig. 7.17) can be created for $MP_T Y_N$ Moore NFSM.

Int this model, the LUTer1 implements the system (2.19). The LUTer2 generates functions $z_r \in Z$ used for encoding the inputs of NLCS $\beta_g \in C_N$. It is the following system:

$$Z = Z(T, P). \tag{7.36}$$

The LUTer3 implements functions (7.11). The EMBer implements functions $y_t(T)$ and $Y(T)$.

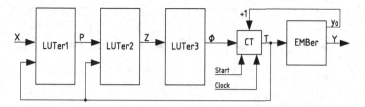

Fig. 7.17 Structural diagram of FPGA-based $MP_T Y_N$ Moore NFSM

Table 7.19 Table of LUTer2 of Moore NFSM $MP_T Y_N(\Gamma_{13})$

a_m	$K(a_m)$	a_s	$K(a_s)$	P_h	Z_h	h
a_3	0010	a_4	001	p_1	z_3	1
		a_7	011	$\bar{p}_1 p_2$	$z_2 z_3$	2
		a_9	100	$\bar{p}_1 \bar{p}_2$	z_1	3
a_6	0110	a_{11}	110	p_2	$z_1 z_2$	4
		a_5	001	$\bar{p}_2$	z_3	5
a_8	1001	a_{12}	111	$p_1 \bar{p}_2$	$z_1 z_2 z_3$	6
		a_{10}	101	$p_1 p_2$	$z_1 z_3$	7
		a_{11}	110	$\bar{p}_1$	$z_1 z_2$	8
a_{10}	1011	a_{12}	111	$p_1 \bar{p}_2$	$z_1 z_2 z_3$	9
		a_{10}	101	$p_1 p_2$	$z_1 z_3$	10
		a_{11}	110	$\bar{p}_1$	$z_1 z_2$	11

The design methods are practically identical for NFSMs MPY_N and $MP_T Y_N$. But there is an additional step in the second case. It is the step connected with constructing the table of LUTer3. This table is the same as the table of LUTer2 for $P_T Y_N$ Moore NFSM.

Let us construct the table of LUTer2 for Moore NFSM $MP_T Y_N(\Gamma_{13})$. There is the following set of inputs $I_N(\Gamma_{13}) = \{a_1, a_4, a_5, a_7, a_9, a_{10}, a_{11}, a_{12}\}$. There is $IN = 8$. Using (7.10), we can find the value $R_{IN} = 3$ and the set $Z = \{z_1, z_2, z_3\}$.

Let us encode the inputs $a_m \in I_N$ in the trivial way: $K(a_1) = 000$, $K(a_4) = 001, \ldots, K(a_{12}) = 111$. The table of LUTer2 has the same columns as its counterpart for the MPY_N Moore NFSM. But there are two differences. Firstly, the column $K(a_s)$ includes the codes of inputs. Secondly, there is a column Z_h instead of Φ_h. In the discussed case, it is Table 7.19.

This table is used for deriving the system (7.36). For example, the following minimized equation can be derived from Table 7.19:

$$z_1 = \bar{T}_1 \bar{T}_2 T_3 \bar{T}_4 \bar{p}_1 \bar{p}_2 \vee \bar{T}_1 T_2 T_3 \bar{T}_4 \bar{p}_2 \vee T_1 \bar{T}_2 T_4. \tag{7.37}$$

The method of RLC can be used together with the optimal natural state assignment and with encoding of the classes of pseudoeqivalent NLCSs. The first combination leads to $MP_0 Y_N$ Moore NFSM. There are two possible models in the second case.

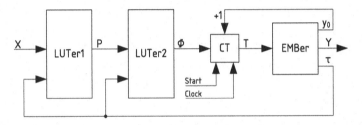

Fig. 7.18 Structural diagram of FPGA-based $MP_{C1}Y_N$ Moore NFSM

They are the models $MP_{C1}Y_N$ and $MP_{C2}Y_N$. Of course, the BIT can be used in these FSMs, too. It leads to the following models $MP_{T0}Y_N$, $MP_{TC1}Y_N$ and $MP_{TC2}Y_N$. For example, let us discuss the $MP_{C1}Y_N$ Moore NFSM (Fig. 7.18).

In this model, the LUTer1 implement the functions

$$P = P(\tau, X). \tag{7.38}$$

The LUTer2 implements the functions

$$\Phi = \Phi(\tau, P). \tag{7.39}$$

The purpose of EMBer is the same as for $P_{C1}Y_N$ Moore NFSM. The design method includes the following steps for $MP_{C1}Y_N$ Moore NFSM:

1. Constructing the set of states A.
2. Constructing the set of NLCSs C_N.
3. Executing the natural state assignment.
4. Constructing the partition Π_{CN}.
5. Executing the encoding of the classes $B_i \in \Pi_{CN}$.
6. Finding the set of additional variables P.
7. Constructing the table of LUTer1.
8. Constructing the table of LUTer2.
9. Constructing the table of EMBer.
10. Implementing the FSM logic circuit.

Let us discuss an example of design for Moore NFSM $MP_{C1}Y_N(\Gamma_{13})$. The first five steps of the design method are already executed (see Table 7.15). The step 6 gives the set $P = \{p_1, p_2\}$.

The table of LUTer1 is constructed using the table of RLC. This table is the same as for MPY_N Moore NFSM but the states $a_m \in B_i$ are replaced by the classes $B_i \in \Pi_{CM}$. It is Table 7.20 in the discussed case.

The table of LUTer1 has the columns B_i, $K(B_i)$, $X(p_1)$, $X(p_2)$, ..., $X(p_G)$, i. In the discussed case, it is Table 7.21.

Table 7.20 Table of RLC for Moore NFSM $MP_{C1}Y_N(\Gamma_{13})$

$p_g \backslash B_i$	B_1	B_2	B_3	B_4
p_1	x_1	–	x_3	–
p_2	x_2	x_2	x_4	–

Table 7.21 Table of LUTer1 for Moore NFSM $MP_{C1}Y_N(\Gamma_{13})$

B_i	$K(B_i)$	$X(p_1)$	$X(p_2)$	i
B_1	00	x_1	x_2	1
B_2	01	–	x_2	2
B_3	10	x_3	x_4	3

Table 7.22 Table of LUTer2 for Moore NFSM $MP_{C1}Y_N(\Gamma_{13})$

B_i	$K(B_i)$	a_s	$K(a_s)$	P_h	Φ_h	h
B_1	00	a_4	0100	p_1	D_2	1
		a_7	1000	$\bar{p}_1 p_2$	D_1	2
		a_9	1010	$\bar{p}_1 \bar{p}_2$	$D_1 D_3$	3
B_2	01	a_{11}	1010	p_2	$D_1 D_2$	4
		a_5	0101	$\bar{p}_2$	$D_2 D_4$	5
B_3	10	a_{12}	1101	$p_1 \bar{p}_2$	$D_1 D_2 D_4$	6
		a_{10}	1011	$p_1 p_2$	$D_1 D_3 D_4$	7
		a_{11}	1100	$\bar{p}_1$	$D_1 D_2$	8

Using the code 11, we can get the following equations from Table 7.21:

$$p_1 = \bar{\tau}_1 x_1 \vee \tau_1 x_3;$$
$$p_2 = \bar{\tau}_1 x_2 \vee \tau_1 x_4. \tag{7.40}$$

Comparison of (7.34) and (7.40) shows that these systems have the same amount of terms and literals. It means that, in the discussed case, the hardware amount is equal for blocks LUTer1 of both NFSMs.

The table of LUTer2 includes the columns B_i, $K(B_i)$, a_s, $K(a_s)$, P_h, Φ_h, h. In the discussed case, the system (7.13) is used for creating Table 7.22.

This table is used for obtaining the system (7.39). For example, there are the following equations derived from Table 7.22:

$$D_1 = \bar{\tau}_1 \bar{\tau}_2 \bar{p}_1 \vee \bar{\tau}_1 \tau_2 \vee \tau_1 \bar{\tau}_2;$$
$$D_4 = \bar{\tau}_1 \tau_2 \bar{p}_2 \vee \tau_1 \bar{\tau}_2 p_1. \tag{7.41}$$

As it is for all previous cases discussed for GSA Γ_{13}, the EMBer includes 16 cells. The code $K(B_i)$ is placed in the cell having address $K(a_m)$ where $a_m = O_g$

Table 7.23 The part of table of EMBer for Moore NFSM $MP_{C1}Y_N(\Gamma_{13})$

$K(a_m) T_1 \ldots T_4$	$Y(a_m) y_0 \ldots y_5$	$K(B_i) \tau_1\tau_2$	h	m
0000	10 0000	00	1	1
0001	11 1000	00	2	2
0010	00 0110	00	3	3
0011	00 0000	00	4	*
0100	10 1100	00	5	4
0101	11 0010	00	6	5
0110	00 0001	01	7	6
0111	00 0000	00	8	*

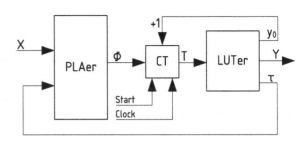

Fig. 7.19 Structural diagram of HFPGA-based $P_{HC}Y_N$ Moore NFSM

and $\beta_g \in B_i$. There is a part of table of EMBer for the discussed case represented by Table 7.23.

The state codes (Fig. 7.16) are used for constructing both Tables 7.22 and 7.23. As in the previous case, we do not discuss the final step of design. Also, we do not discuss other NFSMs having the block of RLC. We hope that our reader has enough information to do it by himself or herself.

7.4   Design of Moore NFSMs with HFPGAs

Obviously, the HFPGAs can be used for implementing logic circuits of NFSMs. If there is no code sharing, then the structure diagrams are the same for both EFSMs and NFSMs. For example, there is the structural diagram of $P_{HC}Y_N$ Moore NFSM shown in Fig. 7.19.

It is obvious that the circuits shown in Figs. 6.20 and 7.19 are identical. There are identical the corresponding design methods. The only difference is reduced to the approach used for the crating LCSs.

Now, let us discuss the design methods targeting NLCS-based FSMs with code sharing. Let us start from $P_{H0}Y_{NC}$ Moore NFSM (Fig. 7.20). Its circuit is represented by the same systems of Boolean functions as it is for PY_{NC} Moore NFSM.

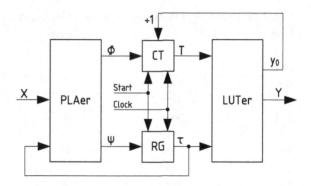

Fig. 7.20 Structural diagram of HFPGA-based $P_{H0}Y_{NC}$ Moore NFSM

There are the following steps in the design method proposed for $P_{H0}Y_{NC}$ Moore NFSM:

1. Creating the set of states A for a given GSA Γ.
2. Constructing the set of NLCSs $C_N = \{\beta_1, \ldots, \beta_{G2}\}$.
3. Constructing the partition $\Pi_{CN} = \{B_1, \ldots, B_{IN}\}$.
4. Executing of the optimal natural state assignment.
5. Constructing the preliminary table of PLAer.
6. Constructing the final table of PLAer.
7. Constructing the table of LUTer.
8. Implementing the FSM logic circuit.

Let us discuss an example of design for the Moore NFSM $P_{H0}Y_{NC}(\Gamma_{14})$. The first four steps are already executed (see Table 7.7). There is the outcome of the optimal natural state assignment shown in Fig. 7.9.

The preliminary table of PLAer is constructed using the system of GFT. It is the system (7.19) for the discussed case. This table is the same as Table 7.10.

The final table of PLAer includes the following columns: *Inputs, Terms, Outputs*. It is constructed in the same manner as Table 6.16. In the discussed case the table of PLAer is represented by Table 7.24. We hope the connection is obvious between Tables 7.10 and 7.24.

The table of LUTer is constructed in the same way as table of EMBer for P_0Y_{NC} Moore NFSM. The part of this table is shown in Table 7.25. It is based on state codes $K(a_m)$ from Fig. 7.9.

Let the following conditions be true for the HFPGA chip used for implementation of the logic circuit of $P_{H0}Y_{CN}(\Gamma_{14})$:

$$s_p \geq L + R_{G2}; \tag{7.42}$$

$$t_p \geq R_{G2} + R_{C2}; \tag{7.43}$$

$$q_p \geq H_E; \tag{7.44}$$

$$s_l \geq R_{G2} + R_{C2}. \tag{7.45}$$

Table 7.24 Final table of PLAer for Moore NFSM $P_{H0}Y_{NC}(\Gamma_{14})$

Inputs		Terms	Outputs
$\tau_1\tau_2\tau_3$	$x_1\ldots x_4$		$D_1D_2D_3D_5$
0*0	1*1*	1	1100
0*0	1*0*	2	1101
0*0	01**	3	1110
0*0	00**	4	1010
1**	**1*	5	1000
1**	**01	6	1100
1**	1*00	7	0100
1**	0*00	8	0010
0*1	**1*	9	0101
0*1	**0*	10	1011

Table 7.25 The part of table of LUTer for Moore NFSM $P_{H0}Y_{NC}(\Gamma_{14})$

$K(a_m)\ \tau_1\tau_2\tau_3T_1T_2$	$Y(a_m)\ y_0\ldots y_7$	h	m
000 00	1000 0000	1	1
000 01	0110 0000	2	2
000 10	0000 0000	3	*
000 11	0000 0000	4	*
001 00	1110 0000	5	15
001 01	1001 0001	6	16
001 10	0000 1000	7	17
001 11	0000 0000	8	*

In this case, only a single PLA macrocell is used for implementing the circuit of PLAer. Also, only 8 LUTs are used for implementing the circuit of LUTer. It leads to the logic circuit shown in Fig. 7.21.

If conditions (7.42)–(7.44) are violated, the more than one macrocell is necessary to implement the circuit of PLAer. To minimize the hardware amount in the PLAer, it is necessary to use the known methods [2, 3]. Of course, these methods should be tuned to meet the peculiarities of NLCS-based Moore FSMs.

Let us discuss the case when the condition (7.42) is violated. Let the following condition (7.46) be true:

$$S_p \geq G + R_{G2}. \tag{7.46}$$

In this case, we can use the method of RLC. Let us denote as DL_2 the difference

$$\Delta_{L2} = S_p + (G = R_{G2}). \tag{7.47}$$

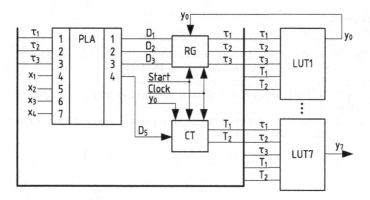

Fig. 7.21 Logic circuit of Moore NFSM $P_{H0}Y_{CN}(\Gamma_{14})$

Fig. 7.22 Structural diagram
of HFPGA-based $MP_H Y_{NC}$

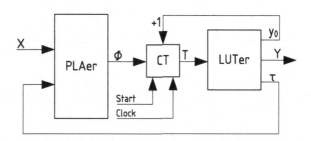

Obviously, the DL_2 inputs of PLA can be used for entering the logical conditions. It means that only DL_1 of logical conditions should be replaced. Let the following condition take place:

$$S_p \geq \Delta_{L1} + R_{G2}. \qquad (7.48)$$

In this case, we propose the following structural diagram of $MP_H Y_{NC}$ Moore NFSM (Fig. 7.22).

In this model, the PLAer1 implements the system

$$P = P(\tau, X^1). \qquad (7.49)$$

It transforms the logical conditions $x_e \in X^1$ into the additional variables $p_g \in P$. There are exactly Δ_{L1} elements in the set X^1. The PLAer2 implements the functions

$$\Psi = \Psi(\tau, x^2, P); \qquad (7.50)$$

$$\Phi = \Phi(\tau, x^2, P). \qquad (7.51)$$

This very approach can be used for optimizing HFPGA based Moore EFSMs. It can be used for all models of NFSMs discussed in this Chapter. We hope that our reader will be able to generate the models of FSMs as well as the corresponding design methods.

References

1. Baranov, S.I.: Logic synthesis for control automata. Kluwer Academic Publishers (1994)
2. Barkalov, A., Titarenko, L.: Logic synthesis for FSM-based control units. Lecture Notes in Electrical Engineering, vol. 53. Springer, Berlin (2009)
3. Barkalov, A., Węgrzyn, M.: Design of control units with programmable logic. UZ Press, Zielona Góra (2006)
4. Wiśniewski, R.: Synthesis of Compositional Microprogram Control Units for Programmable Devices. UZ Press, Zielona Góra (2009)

Chapter 8
Hardware Reduction for Moore XFSMs

Abstract The Chapter is devoted to hardware reduction targeting the extended LCS-based Moore FSMs. Firstly, the design method is proposed for the base model of XFSM. Next, the methods are proposed targeting the hardware reduction in the circuits based on this model. They are based on the executing either optimal state assignment or transformation of state codes. The third part deals with the models based on the encoding of the chain outputs. At last, the principle of code sharing is discussed. In this case, the state code is represented as a concatenation of the code of class of pseudoequivalent chains and the code of element inside this class.

8.1 Design of XFSM with Base Structure

There are three main blocks in the base model of XFSM: the block of input memory functions, the counter and the block of microoperations (see Fig. 3.16). As in the previous cases, the BIMF is implemented as the LUTer, whereas the BMO is implemented as EMBer. Let us start from the model P_1Y_X (Fig. 8.1).

The LUTer implements the functions (2.9) and the following function:

$$y_L = y_L(T, X). \tag{8.1}$$

The EMBer implements the functions (2.17). Let us point out that this model has never been discussed in the literature.

There are the following steps in the proposed design method targeting P_1Y_X Moore XFSMs:

1. Creating the set A for a given GSA Γ.
2. Constructing the set of XLCSs $C_X = \{\gamma_1, \ldots, \gamma_{G3}\}$.
3. Executing the natural state assignment.
4. Constructing the table of LUTer.
5. Constructing the table of EMBer.
6. Implementing the FSM logic circuit.

© Springer International Publishing AG 2018
A. Barkalov et al., *Logic Synthesis for Finite State Machines Based on Linear Chains of States*, Studies in Systems, Decision and Control 113,
DOI 10.1007/978-3-319-59837-6_8

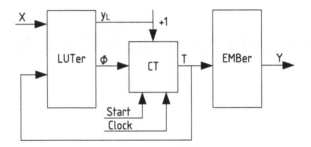

Fig. 8.1 Structural diagram of FPGA-based P_1Y_X XFSM

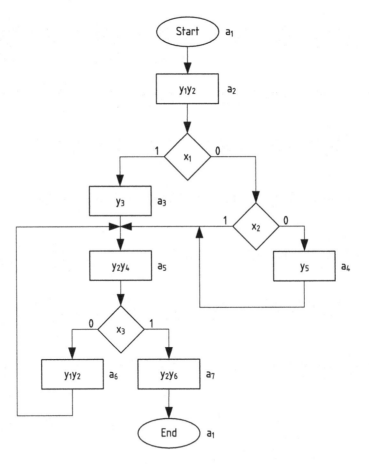

Fig. 8.2 Initial graph-scheme of algorithm Γ_{15}

Let us discuss an example of design for the Moore XFSM $P_1Y_X(\Gamma_{15})$, where the GSA Γ_{15} is shown fi Fig. 8.2.

The following sets can be derived from the GSA Γ_{15}: $A = \{a_1, \ldots, a_7\}$, $X = \{x_1, x_2, x_3\}$, $Y = \{T_1, T_2, T_3\}$ and $\Phi = \{D_1, D_2, D_3\}$. Using the procedure P_4, the

Fig. 8.3 State codes for
Moore XFSM $P_1 Y_X(\Gamma_{15})$

T_3\\$T_1 T_2$	00	01	11	10
0	a_1	a_3	a_6	a_7
1	a_2	a_5	✶	a_4

following set C_X can be found $C_X = \{\beta_1, \beta_2, \beta_3\}$. There are the following XLCS
$\beta_g \in C_X$: $\beta_1 = \langle a_1, a_2, a_3, a_5, a_7 \rangle$, $\beta_2 = \langle a_4 \rangle$, $\beta_3 = \langle a_6 \rangle$.

Let us execute the step 3. The state codes should obey to the relation (3.26). At
the same time, it should be zero code assigned for the initial state $a_1 \in A$ [2]. Using
these rules, the following state codes can be obtained (Fig. 8.3).

The table of LUTer includes the following columns: a_m, $K(a_m)$, a_s, $K(a_s)$, X_h,
Φ_h, y_L, h. The column y_L includes one if the transition $\langle a_m, a_s \rangle$ should satisfy to the
relation $K(a_s) = K(a_m) + 1$. In the discussed case, there is the following table of
LUTer (Table 8.1).

The table of EMBer includes the columns $K(a_m)$, $Y(a_m)$, m. It is constructed as
for all cases discussed before. It is Table 8.2 for the case of $P_1 Y_X(\Gamma_{15})$.

The table of LUTer includes 10 rows. It is the same number as for the Moore FSM
$PY(\Gamma_{15})$. But there are some input memory functions only in five rows of Table 8.1.
Let us point out that each function D_r can include up to 9 terms in the case of
$PY(\Gamma_{15})$. So, we can expect that the logic circuit is simpler for LUTer of $P_1 Y_X(\Gamma_{15})$
than for its counterpart from $PY(\Gamma_{15})$.

The following system can be derived from Table 8.1:

$$
\begin{aligned}
D_1 &= \bar{T}_1 \bar{T}_2 T_3 \bar{x}_1 \bar{x}_2 \vee \bar{T}_1 T_2 T_3 \bar{x}_3 = F_1 \vee F_8; \\
D_2 &= \bar{T}_1 \bar{T}_2 T_3 \bar{x}_1 x_2 \vee \bar{T}_1 \bar{T}_2 \bar{T}_3 \vee \bar{T}_1 T_2 T_3 \bar{x}_3 \vee T_1 T_2 \bar{T}_3 = F_3 \vee F_6 \vee F_8 \vee F_9; \\
D_3 &= \bar{T}_1 \bar{T}_2 T_3 \bar{x}_1 \vee T_1 \bar{T}_2 T_3 \vee T_1 T_2 \bar{T}_3 = [F_3 \vee F_4] \vee F_6 \vee F_9; \\
y_l &= F_1 \vee F_2 \vee F_5 \vee F_7.
\end{aligned}
\tag{8.2}
$$

Table 8.1 Table of LUTer for Moore XFSM $P_1 Y_X(\Gamma_{15})$

a_m	$K(a_m)$	a_s	$K(a_s)$	X_h	Φ_h	y_L	h
a_1	000	a_2	001	1	–	1	1
a_2	001	a_3	010	x_1	–	1	2
		a_5	011	$\bar{x}_1 x_2$	$D_2 D_3$	–	3
		a_4	101	$\bar{x}_1 \bar{x}_2$	$D_1 D_3$	–	4
a_3	010	a_5	011	1	–	1	5
a_4	101	a_5	011	1	$D_2 D_3$	–	6
a_5	011	a_7	100	x_3	–	1	7
		a_6	110	$\bar{x}_3$	$D_1 D_2$	–	8
a_6	110	a_5	011	1	$D_2 D_3$	–	9
a_7	100	a_a	000	1	–	–	10

Table 8.2 Table of EMBer
for Moore XFSM $P_1Y_X(\Gamma_{15})$

$K(a_m)\ T_1T_2T_3$	$Y(a_m)\ y_1 \ldots y_6$	m
000	000000	1
001	110000	2
010	001000	3
011	010100	5
100	010001	7
101	000100	4
110	110000	6
111	000000	*

The expression $[F_3 \vee F_4]$ means that the corresponding term is obtained by using the law of expansion [3]. The equation for y_L could be simplified.

Analysis of Table 8.1 shows that $y_L = 1$ in two cases:

1. $D_1 \vee D_2 \vee D_3 = 0$;
2. $K(a_m) \neq 100$.

It gives the following equations:

$$y_L = \overline{D_1 \vee D_2 \vee D_3} \cdot \overline{T_1\bar{T}_2\bar{T}_3} = \bar{D}_1\bar{D}_2\bar{D}_3\bar{T}_1 \vee \bar{D}_1\bar{D}_2\bar{D}_3T_3 \vee \bar{D}_1\bar{D}_2\bar{D}_3T_3. \quad (8.3)$$

The Eq. (8.3) means that the function y_L can be expressed as:

$$y_L = y_L(\Phi, T). \qquad (8.4)$$

When an FSM is designed, either (8.1) or (8.4) could be chosen for implementing the corresponding circuit. Of course, it is reasonable to choose equation leading to the circuit with less amount of LUTs.

Let us point out that the pulse *Start* should be connected with the clearing input of the counter. Also, the functions $C_1 = y_L Clock$ and $C_2 = \bar{y}_L Clock$ should be implemented by the LUTer.

Let LUTs having $S = 6$ inputs be used for implementing the circuit of LUTer. In this case, functions (8.2), C_1 and C_2 are implemented using a single LUT. Let the FPGA in use include EMBs having the configuration 8×8 (bits). It means that only one EMB is necessary for implementing the circuit of EMBer. The logic circuit of $P_1Y_X(\Gamma_{15})$ is shown in Fig. 8.4.

In this circuit, LUT4 implements the function y_L represented by (8.3), LUT5 implements Eq. (8.5),

$$C_1 = y_1 Clock; \qquad (8.5)$$
$$C_2 = \bar{y}_1 Clock. \qquad (8.6)$$

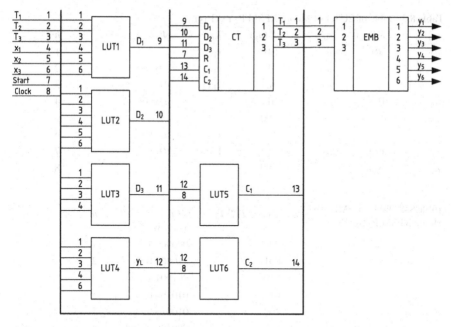

Fig. 8.4 Logic circuit of Moore XFSM $P_1 Y_X (\Gamma_{15})$

Fig. 8.5 Structure diagram of $P_2 Y_X$ Moore FSM

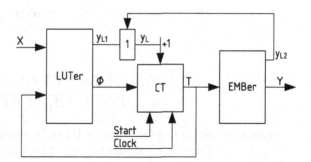

Let the LUTer generate a function y_{L1} initializing the incrementing the counter for conditional transitions. Let the EMBer generates a function y_{L2} initializing the incrementing the counter for unconditional transitions. In this case, the following equation should by implemented:

$$y_L = y_{L1} \vee y_{L2}. \tag{8.7}$$

It leads to $P_2 Y_X$ Moore XFSM shown in Fig. 8.5.

In the case of FSM $P_2 Y_X (\Gamma_{15})$, the LUTer is represented by Table 8.3.

Comparison of Tables 8.1 and 8.3 shows that the later includes less rows. Two different equations could be formed for the function y_{L1}:

Table 8.3 Table of LUTer of Moore XFSM $P_2Y_X(\Gamma_{15})$

a_m	$K(a_m)$	a_s	$K(a_s)$	X_h	Φ_h	y_{L1}	h
a_2	001	a_3	010	x_1	–	1	1
		a_5	011	$\bar{x}_1 x_2$	$D_2 D_3$	–	2
		a_4	101	$\bar{x}_1 \bar{x}_2$	$D_1 D_3$	–	3
a_4	101	a_5	011	1	$D_2 D_3$	–	4
a_5	011	a_7	100	x_3	–	1	5
		a_6	110	$\bar{x}_3$	$D_1 D_2$	–	6
a_6	110	a_5	011	1	$D_2 D_3$	–	7
a_7	100	a_1	000	–	–	–	8

Table 8.4 Table of EMBer of Moore XFSM $P_2Y_X(\Gamma_{15})$

$K(a_m)$ $T_1 T_2 T_3$	$Y(a_m)$ $y_1 \ldots y_6$ y_{L2}	m
000	000000 1	1
001	110000 0	2
010	001001 1	3
011	010100 0	5
100	010010 0	7
101	000100 0	4
110	110000 0	6
111	000000 0	*

$$y_{L1} = \bar{T}_1 \bar{T}_2 T_3 x_1 \vee \bar{T}_1 T_2 T_3 x_3; \qquad (8.8)$$

$$y_{L1} = \overline{D_1 \vee D_2 \vee D_3} \wedge \overline{T_1 \bar{T}_2 \bar{T}_3}. \qquad (8.9)$$

Obviously, the Eq. (8.9) is a part of Eq. (8.2), whereas the Eq. (8.5) coincides with (8.3). The function y_{L2} should be formed for states $a_1, a_3 \in A$. It leads to the following table of EMBer (Table 8.4).

Let V_0 be a number of cells of EMB if it has a single output ($t_F = 1$). To be implemented as single EMB, the following conditions should take places for EMBers of PY_C and PY_{C1} FSMs:

$$2^R N \leq V_0; \qquad (8.10)$$

$$2^R (N + 1) \leq V_0. \qquad (8.11)$$

Analysis of the benchmarks from [4] shows that both conditions (8.10)–(8.11) are satisfied for all benchmarks. So, both models can be used. The criterion of choice could be either minimum hardware or minimum propagation time for function y_L. In this Chapter, we always use the models where the function y_L is generated by the LUTer.

8.2 Optimization of XFSM with the Base Structure

It is known that the Moore FSM can be optimized by using the existence of pseudoe-quivalent states [2]. But it is necessary to change the definition of PES for XLCS-based More FSMs. Let us define PEWS as the following. The states a_m, $a_s \in A$ are pseudoequivalent, if they:

1. belong to different XLCS $\beta_g \in C_x$;
2. mark operator vertices connected with the input of the same vertex of GSA Γ;
3. the condition (3.26) takes no place for transitions from these states.

There are G3 $= 3$ extended LCSs in the case of GSA Γ_{15}. Let Π_{CX} be a partition of the set A by the classes of PES for an XLCS-based Moore FSM. In the discussed case, there is $\Pi_{CX} = \{B_1, \ldots, B_5\}$, where $B_1 = \{a_1\}$, $B_2 = \{a_2\}$, $B_3 = \{a_3\}$, $B_4 = \{a_4, a_6\}$, $B_5 = \{a_5\}$ and $B_6 = \{a_7\}$. So the partition Π_{CX} includes $I_{CX} = 6$ classes of PES. Let us point out that the following relation takes place:

$$I_{CX} \geq I. \tag{8.12}$$

In (8.12), the symbol I stands for the capital numer of the partition Π_A formed for ULCS-based Moore FSM.

Let us encode the PES $a_m \in B_i$ in such a way that:

1. the condition (3.26) takes place for each XLCS $\beta_g \in C_x$;
2. the codes of states $a_m \in B_i$ belong to the same generalized interval of R-dimensional Boolean space.

As in previous cases, it is necessary to execute the optimal natural state assignment. There is a variant of the optimal natural state assignment for the discussed example (Fig. 8.6). Let us point out that the symbol P_0 means that the optimal state encoding is used in an FSM.

Using the "don't care" input assignment 110, the following codes can be found for classes $B_i \in \Pi_{EX}$: $K(B_1) = 000$, $K(B_2) = 001$, $K(B_3) = *10$, $K(B_4) = 1*1$, $K(B_5) = 011$, and $K(B_6) = 1*0$.

There are the following steps in the proposed design method for $P_{01}Y_X$ Moore XFSM:

1. Constructing the set of states A.
2. Constructing the partition $\Pi_{CX} = \{\gamma_1, \ldots, \gamma_{G3}\}$.

Fig. 8.6 Optimal state codes for $P_{01}Y_X(\Gamma_{15})$

T_3 \ T_1T_2	00	01	11	10
0	a_1	a_2	a_5	a_3
1	a_7	a_4	a_6	✳

3. Finding the partition $\Pi_{CX} = \{B_1, \ldots, B_{IC_X}\}$.
4. Executing the optimal natural state assignment.
5. Constructing the table of LUTer.
6. Constructing the table of EMBer.
7. Implementing the FSM logic circuit.

Let us discuss an example of design for the Moore XFSM $P_{01} Y_X (\Gamma_{15})$. The first 4 steps are already executed. The table of LUTer includes the following columns: B_i, $K(B_i)$, a_s, $K(a_s)$, Φ_h, y_L, h. It is constructed on the base of GFTs. In the discussed case, there is the following system of GFTs:

$$
\begin{aligned}
B_1 &\rightarrow a_2; \\
B_2 &\rightarrow x_1 a_3 \vee \bar{x}_1 x_2 a_5 \vee \bar{x}_1 \bar{x}_2 a_4; \\
B_3 &\rightarrow a_5; \\
B_4 &\rightarrow a_5; \\
B_5 &\rightarrow x_3 a_7 \vee \bar{x}_3 a_5; \\
B_6 &\rightarrow a_1.
\end{aligned}
\qquad (8.13)
$$

The table of LUTer includes 9 rows for the discussed case (Table 8.5). The columns Φ_h and y_L are filled using the following approach. Let us consider the transitions from the class $B_i \in \Pi_{EX}$, where $a_m \in B_i$. If the condition (3.26) takes place for the transition $\langle a_m, a_s \rangle$, then $\Phi_h = \emptyset$ and $y_L = 1$. If the condition (3.26) is violated, then $y_L = 0$ and variables $D_r \in \Phi_h$ are determined by the code $K(a_s)$.

The table of EMBer is constructed in the same way as in the previous case. We do not discuss this step.

There is the second known optimization approach, namely the transformation of state codes. In this case, the classes of PES are encoded using the class codes $K(B_i)$ having R_{IX} bits:

$$
R_{IX} = \lceil \log_2 IX \rceil. \qquad (8.14)
$$

Table 8.5 Table of LUTer of Moore NFSM $P_{01} Y_X (\Gamma_{15})$

B_i	$K(B_i)$	a_s	$K(a_s)$	X_h	Φ_h	y_L	h
B_1	000	a_2	001	1	–	1	1
B_2	001	a_3	010	x_1	–	1	2
		a_5	011	$\bar{x}_1 x_2$	$D_2 D_3$	–	3
		a_4	101	$\bar{x}_1 \bar{x}_2$	$D_1 D_3$	–	4
B_3	*10	a_5	011	1	–	1	5
B_4	1*1	a_5	011	1	$D_2 D_3$	–	6
B_5	011	a_7	100	x_3	–	1	7
		a_5	011	$\bar{x}_3$	$D_2 D_3$	–	8
B_6	1*0	a_1	000	1	–	–	9

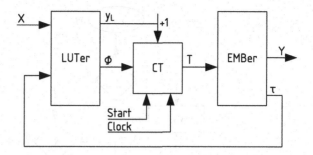

Fig. 8.7 Structural diagram of FPGA-based $P_{1C1}Y_X$ Moore XFSM

There are additional variables $\tau_r \in \tau$ used for encoding of the classes $B_i \in \Pi_{CX}$. Obviously, there is $R_{IX} = |\tau|$. If there are R_{IX} free outputs in EMBs, then the model $P_{1C1}Y_X$ is used (Fig. 8.7).

The block LUTer implements the system (3.57) together with the function

$$y_L = y_L(\tau, X). \tag{8.15}$$

The EMBer generates the functions (2.16) and (3.17).

Let the following conditions take places:

$$t_F > N; \tag{8.16}$$

$$t_F < N + R_{IX}. \tag{8.17}$$

In this case, the model $P_{1C2}Y_X$ can be used (Fig. 8.8).

We hope, our reader understands the functions of each block from Fig. 8.8. Let us discuss the design method for $P_{1C1}Y_X$ Moore XFSM. In includes the following steps:

1. Constructing the set of states A.
2. Constructing the set of XLCSs C_X.
3. Finding the partition Π_{CX}.
4. Executing the natural state assignment.
5. Executing the encoding for classes $B_i \in \Pi_{CX}$.
6. Constructing the table of LUTer.
7. Constructing the table of EMBer.
8. Implementing the FSM logic circuit.

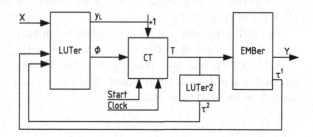

Fig. 8.8 Structural diagram of FPGA-based $P_{1C2}Y_X$ Moore XFSM

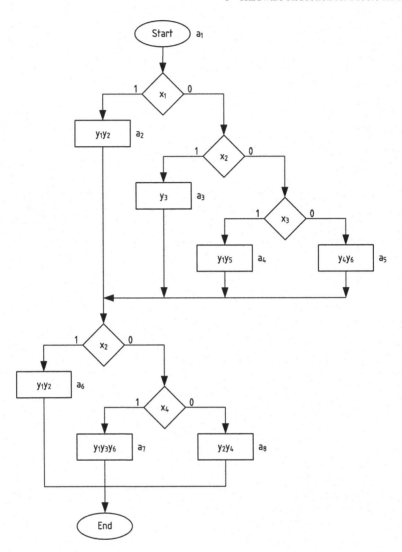

Fig. 8.9 Initial graph-scheme of algorithm Γ_{16}

Let us discuss an example of design for Moore XFSM $P_{1C1}Y_X(\Gamma_{15})$. The initial GSA Γ_{16} is shown in Fig. 8.9.

The following sets and their parameters can be derived from GSA Γ_{16}: $A = \{a_1, \ldots, a_8\}$, $M = 8$, $X = \{x_1, \ldots, x_4\}$, $Y = \{y_1, \ldots, y_6\}$, $N = 6$, $R = 3$, $T = \{T_1, T_2, T_3\}$ and $\Phi = \{D_1, D_2, D_3\}$. The following set of XLCS can be constructed: $C_X = \{\beta_1, \ldots, \beta_6\}$, where $\beta_1 = \langle a_1, a_2, a_6 \rangle$, $\beta_2 = \langle a_3 \rangle$, $\beta_3 = \langle a_4 \rangle$, $\beta_4 = \langle a_5 \rangle$, $\beta_5 = \langle a_7 \rangle$ and $\beta_6 = \langle a_8 \rangle$.

Fig. 8.10 Codes of states of Moore XFSM $P_{1C1}Y_X(\Gamma_{16})$

T_1 \ T_2T_3	00	01	11	10
0	a_1	a_2	a_3	a_6
1	a_4	a_5	a_8	a_7

Using the definition of PES from this Chapter, the set $\Pi_{CX} = \{B_1, \ldots, B_5\}$ can be found. It includes the following classes: $B_1\{a_1\}$, $B_2\{a_2\}$, $B_3\{a_3, a_4, a_5\}$, $B_4\{a_6\}$, $B_5\{a_7, a_8\}$. Because there are classes having more than one state, it is possible to optimize the LUTer. Obviously, there is no variant of the optimal natural state assignment leading to representing each class by a single generalized interval. So, there is sense using the state transformation approach.

Let us encode the states as it is shown in Fig. 8.10.

Using (8.14), we can find $R_{IX} = 3$. Therefore, there is the set $\tau = \{\tau_1, \tau_2, \tau_3\}$. The class assignment can be executed in such a manner that its outcome optimizes the system of input memory functions. Let us start from the system of GFT:

$$B_1 \to x_1 a_2 \vee \bar{x}_1 x_2 a_3 \vee \bar{x}_1 \bar{x}_2 x_3 a_4 \vee \bar{x}_1 \bar{x}_2 \bar{x}_3 a_5;$$
$$B_2 \to x_2 a_6 \vee \bar{x}_2 x_4 a_7 \bar{\bar{x}}_2 \bar{x}_4 a_8;$$
$$B_3 \to x_2 a_6 \vee \bar{x}_2 x_4 a_7 \vee \bar{x}_2 \bar{x}_4 a_8; \qquad (8.18)$$
$$B_4 \to a_1;$$
$$B_5 \to a_1.$$

There are identical GFTs for classes B_2 and B_3 in (8.18). It is possible to simplify the system (3.57) using this fact. To to it, the class codes should be in the same generalized interval of R_{IX}-dimensional Boolean space for the classes B_2 and B_3. The same is true for the classes B_4, $B_5 \in \Pi_{CX}$. One of the variants is shown in Fig. 8.11.

There are the same columns in the table of LUTer for both $P_{01}Y_X$ and $P_{1C1}Y_X$ FSMs. In the discussed case, the table of LUTer is represented by Table 8.6.

In the column $K(B_i)$ the "don't care" class codes are taken into account. The state codes are taken from Fig. 8.10. The system (3.57) can be derived from Table 8.6. After minimizing, this system is the following one:

Fig. 8.11 Optimal class codes for Moore XFSM $P_{1C1}Y_X(\Gamma_{16})$

τ_1 \ $\tau_2\tau_3$	00	01	11	10
0	B_1	B_2	B_3	B_4
1	*	*	*	B_5

Table 8.6 Table of LUTer for Moore XFSM $P_{1C1}Y_X(\Gamma_{16})$

B_i	$K(B_i)$	a_s	$K(a_s)$	X_h	Φ_h	y_L	h
B_1	*00	a_2	001	1	–	1	1
		a_3	011	$\bar{x}_1 x_2$	$D_2 D_3$	–	2
		a_4	100	$\bar{x}_1 \bar{x}_2 x_3$	D_1	–	3
		a_5	101	$\bar{x}_1 \bar{x}_2 \bar{x}_3$	$D_1 D_3$	–	4
B_2	*01	a_6	010	x_2	–	1	5
		a_7	110	$\bar{x}_2 x_4$	$D_1 D_2$	–	6
		a_8	111	$\bar{x}_2 \bar{x}_4$	$D_1 D_2 D_3$	–	7
B_3	*11	a_6	010	x_2	D_2	–	8
		a_7	110	$\bar{x}_2 x_4$	$D_1 D_2$	–	9
		a_8	111	$\bar{x}_2 \bar{x}_4$	$D_1 D_2 D_3$	–	10
B_4	010	a_1	000	1	–	–	11
B_5	110	a_1	000	1	–	–	12

$$D_1 = \bar{\tau}_2 \bar{\tau}_3 \bar{x}_1 \bar{x}_2 \vee \tau_3 \bar{x}_2;$$
$$D_2 = \bar{\tau}_2 \bar{\tau}_3 \bar{x}_1 x_2 \vee \bar{\tau}_2 \tau_3 x_2 \vee \tau_2 \tau_3; \qquad (8.19)$$
$$D_3 = \bar{\tau}_2 \bar{\tau}_3 x_2 \vee \bar{\tau}_2 \bar{\tau}_3 \bar{x}_1 \bar{x}_2 \bar{x}_3 \vee \tau_3 \bar{x}_2 \bar{x}_4.$$

The function y_1 is equal to 1 if $D_1 \vee D_2 \vee D_3 = 0$ (rows 1 and 5) and $B_i = B_1$ of $B_i = B_2$. It gives the following expression: $y_L = \overline{D_1 \vee D_2 \vee D_3} \bar{\tau}_2 = \bar{D}_1 \bar{D}_3 \bar{D}_3 \bar{\tau}_2$. The variable $\tau_2 = 0$ for $K(B_1)$ and $K(B_2)$ (Fig. 8.10).

The table of EMBer includes the columns $K(a_m)$, $Y(a_m)$, $K(B_i)$, m. The column $K(B_i)$ includes the code of a class B_i such that $a_m \in B_i$. In the discussed case, only the variables $\tau_2, \tau_3 \in \tau$ should be shown in this column (Table 8.7).

Obviously, this model can be used if the following condition takes place:

$$t_F \geq N + R_{IX}. \qquad (8.20)$$

Table 8.7 Table of EMBer for Moore XFSM $P_{1C1}Y_X(\Gamma_{16})$

$K(a_m)$ $T_1 T_2 T_3$	$Y(a_m)$ $y_1 \dots y_6$	$K(B_i)$ $\tau_2 \tau_3$	m
000	000000	00	1
001	110000	01	2
010	110000	10	3
011	001000	11	4
100	000010	11	5
101	000101	11	6
110	101001	10	7
111	010100	10	8

To use only a single EMB in the EMBer, the following condition should take place:

$$2^R(N + R_{IX}) \leq V_0. \tag{8.21}$$

In this book, we do not discuss the design methods when the condition (8.22) is violated. As shown our analysis of [4], the condition (8.21) takes place for all benchmarks from the library [4].

8.3   Encoding of Chain Outputs

As it is mentioned before, each XLCS $\gamma_g \in C_X$ can include more than one output. The state $a_m \in A(\gamma_g)$ is an output of the XLCS $\gamma_g \in C_X$ if there is a transition $\langle a_m, a_s \rangle$ such tat $K(a_s) \neq K(a_m) + 1$. These states form the set of outputs $O_X(\Gamma_j)$. For example, there is the set $O_X(\Gamma_{15}) = \{a_2, a_4, a_5, a_6, a_7\}$. Such a set can be found after executing the natural state assignment (3.26).

Analysis of GSA Γ_{15} shows that there are the same transitions for the outputs $a_4, a_6 \in O_X(\Gamma_{15})$. Let us name such outputs pseudoequivalent outputs (PEO). Let us find a partition $\Pi_{XO} = \{O_1, \ldots, O_y\}$ for the set $O_X(\Gamma_j)$. Each element Π_{XO} is a class of PEO. There is the partition $\Pi_{XO} = \{O_1, \ldots, O_4\}$ in the case of Γ_{15}. It includes the following classes: $O_1 = \{a_2\}$, $O_2 = \{a_4, a_6\}$, $O_3 = \{a_5\}$ and $O_4 = \{a_7\}$.

Let us encode the classes $O_j \in \Pi_{XO}$ by binary codes $K(O_j)$ having R_{XO} bits:

$$R_{XO} = \lceil \log_2 J \rceil. \tag{8.22}$$

Let us use the variables $z_r \in Z$ for encoding of the classes $O_j \in \Pi_{XO}$, where $|Z| = R_{XO}$. Now, the model of $P_3 Y_X$ Moore XFSM is proposed (Fig. 8.12).

In this model, the BIMF implements the system of input memory functions

$$\Phi = \Phi(Z, X) \tag{8.23}$$

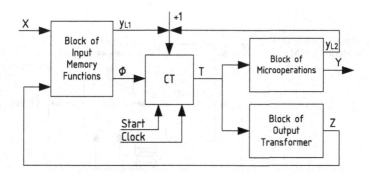

Fig. 8.12 Structural diagram of $P_3 Y_X$ Moore XFSM

together with the function

$$y_{L1} = f_1(Z, X). \tag{8.24}$$

The BMO implements the system of microoperations and the functions

$$y_{L2} = f_2(T). \tag{8.25}$$

The block of output transformer (BOT) implements the functions

$$Z = Z(T). \tag{8.26}$$

If FPGAs are used for implementing an FSM circuit, three different models are possible for XFSM based on the encoding of outputs. Let us discuss them.

Let the following condition take place:

$$t_F \geq N + R_{XO} + 1. \tag{8.27}$$

In this case, the functions (8.25), (8.26) and (2.15) are implemented by the EMBer. It leads to $P_{3C1}Y_X$ Moore XFSM (Fig. 8.13).

Let the following condition take place:

$$\begin{aligned} t_F &> N + 1; \\ t_F &< N + R_{XO} + 1. \end{aligned} \tag{8.28}$$

In this case, the set Z should be divided by two subsets: $Z = Z^1 \cup Z_2$. The functions $z_r \in Z^1$ are generated by the EMBer. Their number t_E is determined as

$$t_E = t_F - (N + 1). \tag{8.29}$$

The functions $z_r \in Z^2$ are implemented by the LUTer2. Their number t_L is determined as

$$t_L = R_{XO} - t_E. \tag{8.30}$$

Fig. 8.13 Structural diagram of $P_{3C1}Y_X$ Moore XFSM

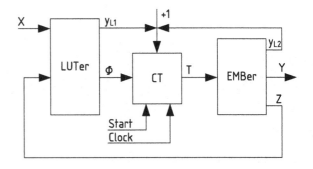

Fig. 8.14 Structural
diagram of FPGA-based
$P_{3C2}Y_X$ Moore XFSM

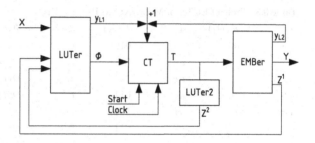

This approach leads to $P_{3C2}Y_X$ Moore XFSM (Fig. 8.14).

Let the following condition take place:

$$t_F = N + 1. \tag{8.31}$$

In this case, the model $P_{3C3}Y_X$ Moore XFSM can be used. In is the same as the one shown in Fig. 8.12. In the case of FPGA, the BOT is represented by the LUTer2.

All these XFSMs are designed in the similar ways. The difference is reduced to the distribution of functions $z_r \in Z$ between the block EMBer and LUTer2. For example, let us discuss the proposed design method for $P_{3C1}Y_X$ Moore XFSM. It includes the following steps:

1. Constructing the set of states A.
2. Constructing the set of XLCS C_X.
3. Natural state assignment.
4. Constructing the set of chain outputs $O_X(\Gamma_j)$.
5. Constructing the partition Π_{XO}.
6. Encoding of the classes of PEO $O_j \in \Pi_{XO}$
7. Constructing the table of LUTer.
8. Constructing the table of EMBer.
9. Implementing the logic circuit of FSM.

Let us discuss an example of design for the Moore XFSM $P_{3C1}Y_X(\Gamma_{15})$. The steps 1–5 are already executed. In the discussed case, there are $R_{XO} = 2$ and $Z = \{z_1, z_2\}$. Let us encode the classes $O_j \in \Pi_{XO}$ in the trivial way: $K(O_1)=00, \ldots, K(O_4)=11$.

To construct the table of LUTer, let us find the system of generalized formula of transitions. In the case of GSA Γ_{15}, this system is the following:

$$
\begin{aligned}
O_1 &\rightarrow x_1 a_3 \vee \bar{x}_1 x_2 a_5 \vee \bar{x}_1 \bar{x}_2 a_4; \\
O_2 &\rightarrow a_5; \\
O_3 &\rightarrow x_3 a_7 \vee \bar{x}_3 a_5.
\end{aligned}
\tag{8.32}
$$

As in all previous cases, D flip-flops are used for implementing the counter. It means that transitions into state $a_1 \in A$ are executed automatically (using only the pulse *Clock*). Due to it, the transitions from the output O_4 are not considered in (8.31).

Table 8.8 Table of LUTer of Moore XFSM $P_{3C1}Y_X(\Gamma_{15})$

O_j	$K(O_j)$	a_s	$K(a_s)$	X_h	Φ_h	y_{L1}	h
O_1	00	a_3	010	x_1	–	1	1
		a_5	011	$\bar{x}_1 x_2$	$D_2 D_3$	–	2
		a_4	101	$\bar{x}_1 \bar{x}_2$	$D_1 D_3$	–	3
O_2	*0	a_5	011	1	$D_2 D_3$	–	4
O_3	1*	a_7	100	x_3	–	1	5
		a_5	011	$\bar{x}_3$	$D_2 D_3$	–	6

The table of LUTer includes the following columns: O_j, $K(O_j)$, a_s, $K(a_s)$, X_h, Φ_h, y_{L1}, h. In the discussed case, the table of LUTer is represented by Table 8.8.

Because the transitions from O_4 are not considered, the code $K(O_4)$ is treated as "don't care". It is used for simplifying the codes of classes O_2 and O_3. The following Boolean functions can be derived from Table 8.8:

$$
\begin{aligned}
D_1 &= \bar{z}_1 \bar{z}_2 \bar{x}_1 \bar{x}_2; \\
D_2 &= \bar{z}_1 \bar{z}_2 \bar{x}_1 x_2 \vee z_1 \vee z_1; \\
D_3 &= \bar{z}_1 \bar{z}_2 \bar{x}_1 \vee z_2 \vee z_1 x_3; \\
y_{L1} &= \bar{z}_1 \bar{z}_2 x_1 \vee z_1 x_3.
\end{aligned}
\tag{8.33}
$$

Analysis of the system (8.33) shows that each its equations can be implemented using a LUT having $S = 4$. This system is much simpler that the system (8.2). Let us point out that the equation $y_{L1} = \bar{D}_3$ can be derived from Table 8.8, as well as $y_{L1} = \bar{D}_1 \bar{z}_1 \vee \bar{D}_2 \bar{z}_1$.

The table of EMBer includes the following columns: $K(a_m)$, $Y(a_m)$ $K(O_j)$, y_{12}, m. In the discussed case, it is represented by Table 8.9.

The logic circuit of Moore XFSM $P_{3C1}Y_X(\Gamma_{15})$ is shown in Fig. 8.15. The function y_L is implemented using the equation $y_L = y_{L1} \vee y_{L2}$. It is implemented by LUTs. Two LUTs (LUT6 and LUT 7) are used for implementing Eqs. (8.5) and (8.6). The EMBer should have the configuration 8×9 bits.

Table 8.9 Table of EMBer of Moore XFSM $P_{3C1}Y_X(\Gamma_{15})$

$K(a_m)$ $T_1 T_2 T_3$	$Y(a_m)$ $y_1 \ldots y_6$	y_{L2}	m	$K(O_j)$ $z_1 z_2$
000	000000	1	1	00
001	110000	0	2	00
010	001000	1	3	00
011	010100	0	5	10
100	010001	0	7	11
101	000010	0	4	01
110	110000	0	6	01
111	000000	0	*	**

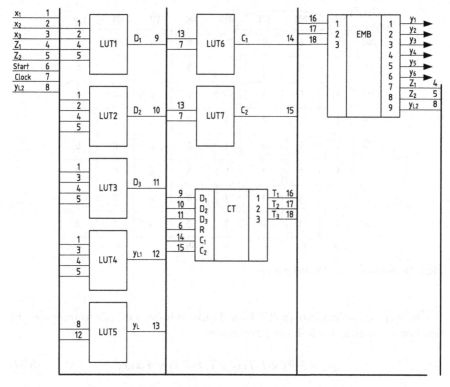

Fig. 8.15 Logic circuit of Moore XFSM $P_{3C1} Y_X(\Gamma_{15})$

Let us point out that the function y_L could be implemented as the following one:

$$y_L = y_L(y_{L1}, T). \qquad (8.34)$$

In the discussed case, the following equation can be found:

$$y_L = y_{L1} \vee A_1 \vee A_3 = y_{L1} \vee \bar{T}_1 \bar{T}_3. \qquad (8.35)$$

It is enough a single LUT having $S = 3$ for implementing the function (8.35). If the following condition takes place

$$R \leq S_L, \qquad (8.36)$$

then circuit of BOT is implemented using only R_0 of LUTs.

The following approach can be used for simplification of the circuit generating the function y_L. Let us discuss it using Table 8.1. Let us construct a Karnaugh map for function y_L (Fig. 8.16) on the base of Table 8.1.

$X_1X_2X_3$ \ $T_1T_2T_3$	000	001	011	010	110	111	101	100
000	1	0	0	1	0	*	0	0
001	1	0	1	1	0	*	0	0
011	1	0	1	1	0	*	0	0
010	1	0	0	1	0	*	0	0
110	1	1	0	1	0	*	0	0
111	1	1	1	1	0	*	0	0
101	1	1	1	1	0	*	0	0
100	1	1	0	1	0	*	0	0
	a_1	a_2	a_5	a_3	a_6		a_4	a_7

Fig. 8.16 Karnaugh map for function y_L

The connection between Table 8.1 and Fig. 8.16 is obvious. Let us minimize the function y_L. It gives the following expression:

$$y_L = \bar{T}_1\bar{T}_3 \vee \bar{T}_2\bar{T}_3 x_1 \vee \bar{T}_1 T_2 \bar{x}_1 x_3 \vee \bar{T}_1 x_1 x_3. \tag{8.37}$$

This expression can be used for the case of encoding of chain outputs. It leads to $P_{3C4}Y_X$ Moore XFSM (Fig. 8.17). In this model, the LUTer2 implements the function $y_L = y_L(T, X)$. In the discussed case, this equation is the Eq. (8.37).

This very approach can be used for optimizing the number of LUTs in P_{1C1} and P_{1C2} Moore XFSMs. For example, the following model can be proposed (Fig. 8.18).

This model is based on the model of $P_{1C1}Y_X$ Moore XFSM (Fig. 8.7). The function (8.36) is used in this model for incrementing the counter CT. We do not discuss the corresponding design method in this Chapter.

Fig. 8.17 Structural diagram of FPGA-based $P_{3C4}Y_X$ Moore XFSM

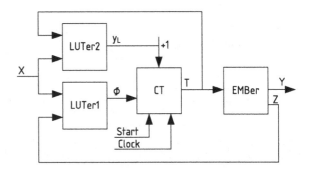

Fig. 8.18 Structural
diagram of FPGA-based
$P_{1C4}Y_X$ Moore XFSM

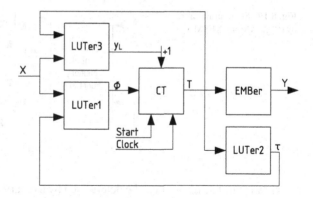

8.4 Code Sharing for Moore XFSMs

Let us find the partition $\Pi_A = \{B_1, \ldots, B_K\}$ of the set A by the classes of PES. We
treat the states as pseudoequivalent if they mark the vertices connected with the input
of the same vertex of GSA Γ [2]. Let us construct the linear chains of classes (LCC)
for a given GSA Γ. Let us define and LCC as a vector $\delta_q = \langle B_{q1}, \ldots, B_{qI_q} \rangle$ where
a transition $\langle a_m, a_s \rangle$ exists for any pair of adjacent components of δ_q. It means that
$a_m \in B_{qi}$ and $a_s \in B_{qi+1}(i = 1, \ldots, I_q - 1)$. A transition could be either conditional
or unconditional.

Let us construct a set $\Pi_\Delta = \{\delta_1, \ldots, \delta_Q\}$ such that each class $B_i \in \Pi_A$ belongs to
a single LCC. So, the set Π_Δ is a partition of Π_A. Let us encode the classes $B_i \in \Pi_A$
by binary codes $K(B_k)$ having R_K bits, where:

$$R_K = \lceil \log_2 K \rceil. \tag{8.38}$$

Let us encode the classes in the natural order:

$$K(B_{qi+1} = K(B_{qi}) + 1. \tag{8.39}$$

In (8.39), there are $q \in \{1, \ldots, Q\}, i \in \{1, \ldots, I_q - 1\}$. Let a class $B_k \in \Pi_A$ include
M_k states $a_m \in A(k = 1, \ldots, K)$. Let us find the value

$$M_0 = \max(M_1, \ldots, M_K). \tag{8.40}$$

Let us encode each state $a_m \in B_k$ by a binary code $C(a_m)$ having R_M bits:

$$R_M = \lceil \log_2 M_0 \rceil. \tag{8.41}$$

Fig. 8.19 Structural diagram
of PY$_{XC}$ Moore XFSM

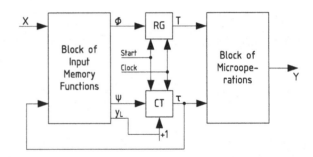

Now each state $a_m \in A$ can be determined by a binary code $K(a_m)$ which is equal
to a concatenation of codes $K(B_k)$ and $C(a_m)$, where $a_m \in B_k$:

$$K(a_m) = K(B_k) * C(a_m). \qquad (8.42)$$

In (8.42), the sign * determines the concatenation of codes.

Let us name such a representation of state as code sharing. Let us use the following
approach for encoding of the states. If there is a transition $\langle a_m, a_s \rangle$ such that $a_m \in B_{qi}$
and $a_s \in B_{qi+1}$, then $C(a_s) = 00 \ldots 0$. It allows to propose the model PY$_{X6}$ Moore
XFSM (Fig. 8.19).

In this FSM, the BIMF implements the functions

$$\Phi = \Phi(\tau, X; \qquad (8.43)$$

$$\Psi = \Psi(\tau, X); \qquad (8.44)$$

$$y_L = y_L(\tau, X). \qquad (8.45)$$

The set Φ includes R_M of functions, whereas the set Ψ R_K of functions. The
register RG keeps the state codes $C(a_m)$ represented by state variables $T_r \in T$, where
$|T| = R_M$. The counter CT keeps the class codes $K(B_k)$ represented by the class
variables $\tau_r \in \tau$, where $|\tau| = R_K$. The variable y_L is used for incrementing the
content of CT. The BMO generates the functions

$$Y = Y(\tau, T). \qquad (8.46)$$

The XFSM operates in the following manner. If *Start=1*, then zero codes are
loaded in both CT and RG. It corresponds to the initial state $a_1 \in A$. If a transition
$\langle a_m, a_s \rangle$ corresponds to (8.39), then the variable y_L is generated, whereas $D_r =
0(D_r \in \Phi \cup \Psi)$. If the condition (8.39) is violated for a given transition, then $y_L = 0$.
In this case contents of RG and CT are determined by functions (8.43)–(8.44).

If FPGAs are used for implementing the logic circuit of PY$_{XC}$ XFSM, then BIMF
is implemented by the LUTer and BMO is implemented by the EMBer (Fig. 8.20).

Fig. 8.20 Structural diagram
of PY$_{XC}$ Moore XFSM
implemented with FPGA

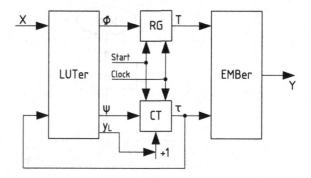

The proposed design method of PY$_{XC}$ Moore XFSM includes the following steps:

1. Constructing the set of internal states A.
2. Constructing the partition $\Pi_A = \{B_1, \ldots, B_K\}$.
3. Constructing the set of LCC $\Pi_B = \{\delta_1, \ldots, \delta_q\}$.
4. Natural encoding of classes $B_k \in \Pi_A$.
5. Encoding of states and finding the codes $K(a_m)$.
6. Constructing the table of LUTer.
7. Constructing the table of EMBer.
8. Implementing the logic circuit of FSM.

Let us discuss an example of design for Moore FSM PY$_{XC}(\Gamma_{17})$ where GSA Γ_{17} is shown in Fig. 8.21.

The following sets and their parameters can be found from GSA Γ_{17}: $A = \{a_1, \ldots, a_{12}\}$, $M = 12$, $R = 4$, $X = \{x_1, \ldots, x_4\}$, $L = 4$, $Y = \{y_1, \ldots, y_6\}$, $N = 6$.

The following partition $\Pi_A = \{B_1, \ldots, B_8\}$ can be found from GSA Γ_{17}. It includes the following classes of PES: $B_1 = \{a_1\}$, $B_2 = \{a_2\}$, $B_3 = \{a_3\}$, $B_4 = \{a_4, a_5, a_6\}$, $B_5 = \{a_7\}$, $B_6 = \{a_8\}$, $B_7 = \{a_9, a_{10}\}$ and $B_8 = \{a_{11}, a_{12}\}$. So, there is $K = 8$; it gives $R_K = 3$ and $\tau = \{\tau_1, \tau_2, \tau_3\}$.

The following set of LCC $\Pi_B = \{\beta_1, \beta_2\}$ can be found from GSA Γ_{17}. It includes the following linear chains of classes: $\delta_1 = \langle B_1, B_2, B_3, B_4, B_5, B_6, B_8 \rangle$ and $\delta_2 = \langle B_7 \rangle$. Let us encode the classes $B_i \in \Pi_A$ in the natural order (Fig. 8.22).

Analysis of the classes $B_i \in \Pi_A$ shows that there are $M_1 = M_2 = M_3 = M_5 = M_6 = 1$, $M_4 = 3$, $M_7 = M_8 = 2$. Therefore, there are $M_0 = 3$ $R_M = 2$ and $T = \{T_1, T_2\}$. The following state codes $C(a_m)$ can be found from GSA Γ_{17}: $C(a_1) = C(a_2) = C(a_3) = C(a_6) = C(a_7) = C(a_8) = C(a_9) = C(a_{11}) = 00$, $C(a_4) = C(a_{10}) = 01$ and $C(a_5) = 10$. Now, the state codes are shown in Fig. 8.23.

To construct the table of LUTer, let us construct the system of generalized formulae of transitions. If the discussed case, it is the following:

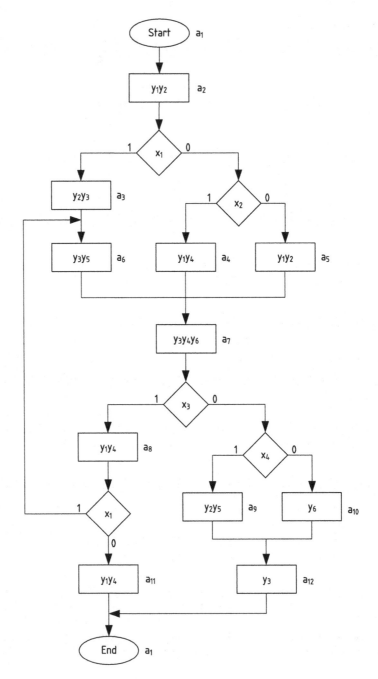

Fig. 8.21 Initial graph-scheme of algorithm Γ_{17}

T_1 \ T_2T_3	00	01	11	10
0	B_1	B_2	B_4	B_3
1	B_5	B_6	B_7	B_8

Fig. 8.22 Codes of classes of XFSM $PY_{XC}(\Gamma_{17})$

T_1T_2 \ $T_1T_2T_3$	000	001	011	010	110	111	101	100
00	a_1	a_2	a_6	a_3	a_{11}	a_9	a_8	a_7
01	*	*	a_4	*	a_{12}	a_{10}	*	*
10	*	*	a_5	*	*	*	*	*
11	*	*	*	*	*	*	*	*
	B_1	B_2	B_4	B_3	B_8	B_7	B_6	B_5

Fig. 8.23 State codes for XFSM $PY_{XC}(\Gamma_{17})$

$$
\begin{aligned}
B_1 &\to a_2; \\
B_2 &\to x_1 a_3 \vee \bar{x}_1 x_2 a_4 \vee \bar{x}_1 \bar{x}_2 a_5; \\
B_3 &\to a_6; \\
B_4 &\to a_7; \\
B_5 &\to x_3 a_8 \vee \bar{x}_3 x_4 a_9 \vee \bar{x}_3 \bar{x}_4 a_{10}; \\
B_6 &\to x_1 a_6 \vee \bar{x}_1 a_{11}; \\
B_7 &\to a_{12}; \\
B_8 &\to a_1.
\end{aligned}
\tag{8.47}
$$

The table of LUTer includes the following columns: B_k, $K(B_k)$, a_s, $K(a_s)$, X_h, Ψ_h, Φ_h, y_1, h. In the discussed case it is represented by Table 8.10.

The table of LUTer is used for deriving the systems (8.43)–(8.44) and the equations for y_L. The following equations can be found from Table 8.10:

$$
\begin{aligned}
D_1 &= \tau_1 \bar{\tau}_2 \bar{\tau}_3 \bar{x}_3 \vee \tau_1 \tau_2 \tau_3; \\
D_2 &= \tau_1 \tau_2 \tau_3 \bar{x}_1 \vee \tau_1 \bar{\tau}_2 \bar{\tau}_3 \bar{x}_3 \vee \tau_1 \bar{\tau}_2 \tau_3 x_1 \vee \tau_1 \tau_2 \tau_3; \\
D_3 &= \bar{\tau}_1 \bar{\tau}_2 \tau_3 \bar{x}_1 \vee \tau_1 \bar{\tau}_2 \bar{\tau}_3 \bar{x}_3 \vee \tau_1 \bar{\tau}_2 \tau_3 x_1; \\
D_4 &= \bar{\tau}_1 \bar{\tau}_2 \tau_3 \bar{x}_1 \bar{x}_2; \\
D_5 &= \bar{\tau}_1 \bar{\tau}_2 \tau_3 \bar{x}_1 x_2 \vee \tau_1 \bar{\tau}_2 \bar{\tau}_3 \bar{x}_3 \bar{x}_4 \vee \tau_1 \tau_2 \tau_3; \\
y_L &= \bar{D}_2 \overline{\tau_1 \tau_2 \tau_3} = \bar{D}_2 \bar{\tau}_1 \vee \bar{D}_2 \bar{\tau}_2 \vee \bar{D}_2 \tau_3.
\end{aligned}
\tag{8.48}
$$

Table 8.10 Table of LUTer of Moore XFSM $PY_{XC}(\Gamma_{17})$

B_k	$K(B_k)$	a_s	$K(a_s)$	X_h	Ψ_h	Φ_h	y_1	h
B_1	000	a_2	00100	1	–	–	1	1
B_2	001	a_3	01000	x_1	–	–	1	2
		a_4	01101	$\bar{x}_1 x_2$	$D_2 D_3$	D_5	–	3
		a_5	01110	$\bar{x}_1 \bar{x}_2$	$D_2 D_3$	D_4	–	4
B_3	010	a_6	01100	1	–	–	1	5
B_4	011	a_7	10000	1	–	–	1	6
B_5	100	a_8	10100	x_3	–	–	1	7
		a_9	11100	$\bar{x}_3 x_4$	$D_1 D_2 D_3$	–	–	8
		a_{10}	11101	$\bar{x}_3 \bar{x}_4$	$D_1 D_2 D_3$	D_5	1	9
B_6	101	a_6	01100	x_1	$D_2 D_3$	–	1	10
		a_{11}	11000	$\bar{x}_1$	–	–	1	11
B_7	111	a_{12}	11001	1	$D_1 D_2$	D_5	–	12
B_8	110	a_1	00000	1	–	–	–	13

The last equation of (8.48) is based on analysis of Table 8.10. The function $D_2 = 1$ if $y_L = 0$. The only exception is the row 13, where $D_2 = y_L = 0$. This row is determined by the conjunction $\tau_1 \tau_2 \bar{\tau}_3$.

The table of EMBer includes the columns $K(a_m)$, $Y(a_m)$, m. In the discussed case it includes 32 rows, because there is $R_K + R_M = 5$. The part of this table for states $a_3, a_4, a_5, a_6 \in A$ is represented by Table 8.11.

The connection between Table 8.11, Figs. 8.21 and 8.23 is obvious. Let us use FPGAs having LUTs with $s = 5$ for implementing the circuit of LUTer. Analysis of the system (8.48) shows that functions D_1, D_2, D_3, D_4, y_L can be implemented using only a single LUT. To implement the function D_5, it should be decomposed:

$$D_5 = \tau_1(\bar{\tau}_2 \bar{\tau}_3 \bar{x}_3 \bar{x}_4 \vee \tau_2 \tau_3) \vee \bar{\tau}_1(\bar{\tau}_2 \tau_3 \bar{x}_1 x_2) = \tau_1 G_1 \vee \bar{\tau}_1 G_2. \qquad (8.49)$$

Table 8.11 Part of table of EMBer for Moore XFSM $PY_{XC}(\Gamma_{17})$

$K(EY_q)\ \tau_1 \tau_2 \tau_3 T_1 T_2$	$Y(a_m)\ y_1 \ldots y_6$	m
010 00	011000	3
010 01	000000	*
010 10	000000	*
010 11	000000	*
011 00	001010	6
011 01	100100	4
011 10	110000	5
011 11	000000	*

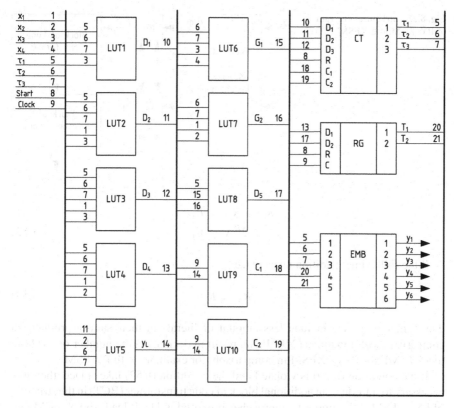

Fig. 8.24 Logic circuit of Moore XFSM $PY_{XC}(\Gamma_{17})$

The logic circuit of Moore XFSM $PY_{XC}(\Gamma_{17})$ is shown in Fig. 8.24.

In this circuit, LUT9 implements the Eq. (8.5), whereas the Eq. (8.6) is implemented by LUT10. If $y_L = 0$, the both CT and RG are loaded using input memory functions. It means that inputs C_2 of CT and C of RG should be connected with the output of LUT10.

8.5 Code Sharing with a Single EMB

To implement the circuit of BMO as as single EMB, the following condition should take place:

$$2^{R_K + R_M} N \le V_0. \tag{8.50}$$

Let only a single EMB could be used in FSM design. All other are used for implementing other parts of a digital system. Let the condition (8.50) is violated, but the following conditions take places:

Fig. 8.25 Structural diagram of PY_{XC1} Moore XFSM

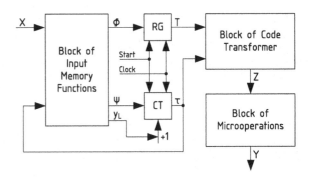

$$R_K + R_M > R; \qquad (8.51)$$

$$2^R N \le V_0. \qquad (8.52)$$

If the following condition takes place

$$R_K < R, \qquad (8.53)$$

then functions (8.43) include less amount of literals in their sum-of-products in comparison with functions (2.9). If can result in decreasing the number of LUTs in block BIMF for PY_{XC} XFSM in comparison, for example, with PY FSM.

If the condition (8.29) is violated and the condition (8.52) takes place, then it is necessary to introduce an additional block of code transformer (BCT) in the structure of PY_{XC} FSM. It is similar to approaches proposed in [1]. It leads to PY_{XC1} Moore XFSM (Fig. 8.25).

In PY_{XC1} Moore XFSM, the BIMF implements functions (8.43)–(8.45). The block BCT implements the functions

$$Z = Z(\tau, T). \qquad (8.54)$$

These functions are used as state variables in PY Moore FSM. The BMO implements the system of microoperations

$$Y = Y(Z). \qquad (8.55)$$

If only a single EMB could be used in the FSM circuit, then BCT is implemented as LUTer1 (Fig. 8.26).

The design method for PY_{XC1} Moore XFSM includes all steps discussed for design pf PY_{XC} XFSM. Also, it includes two following additional steps:

6^a. State assignment.
6^b. Constructing the table of LUTer1.

Let the EMBs in use include configurations 16×8, bits. Let us discuss an example of design for Moore FSM $PY_{CS1}(\Gamma_{17})$. It was found that $R_K = 3$, $R_M = 2$

Fig. 8.26 Structural diagram of PY_{XC1} Moore XFSM implemented with FPGA

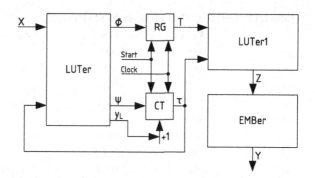

and $R_K + R_M = 5$. It means that the condition (8.50) is violated and the model $PY_{XC}(\Gamma_{17})$ cannot be used. There is the set of states $A = \{a_1, \ldots, a_{12}\}$, therefore, $R = 4$. Because $N = 6$ and $R = 4$, the inequality (8.52) gives the following result: $16 \times 6 \leq 108$. Therefore, the model $PY_{CS1}(\Gamma_{17})$ can be used.

The steps 1–6 are already executed. Let us encode the states $a_m \in A$ by binary codes $CS(a_m)$ having R bits. In the discussed case, there is $R = 4$ and $Z = \{z_1, \ldots, z_4\}$. The outcome of this steps execution does not affect hardware amounts for LUTer and EMB. Let us use the following codes: $CS(a_1) = 0000, CS(a_2) = 0001, \ldots, CS(a_{12}) = 1011$. Now, the step 6^b can be executed.

The table of LUTer1 includes the following columns: a_m, $K(a_m)$, $CS(a_m)$, Z_m, m. The codes $K(a_m)$ are represented in the form (8.42). The column Z_m includes variables $z_r \in Z$ equal to 1 in the code $CS(a_m)$. This table is the base for constructing the system (8.54). The functions of this system are represented as

$$z_r = \bigvee_{m=1}^{M} C_{mr} \left(\bigwedge_{r=1}^{R_K} \tau^{l_{mr}} \right) \left(\bigwedge_{r=1}^{R_M} T^{E_{mr}} \right). \tag{8.56}$$

In (8.56), the Boolean variable $C_{mr} = 1$ iff the bit number r of the code $CS(a_m)$ is equal to 1 ($r = 1, \ldots, R$); the variable l_{mr} is the value of the bit number r of the code $K(B_i)$ where $a_m \in B_i (r = 1, \ldots, R_K)$; the variable E_{mr} is the value of the bit number r of the code $C(a_m)$ where $r = 1, \ldots, R_M$; $l_{mr}, E_{mr} \in \{0, 1, *\}$, $\tau_r^0 = \bar{\tau}_r$, $\tau_r^1 = \tau_r, \tau_r^* = 1$; $T_r^1 = T_r, T_r^0 = \bar{T}_r, T_r^* = 1$.

In the discussed case, this table includes 12 rows (Table 8.12).

Let the following condition take place

$$S_l \geq R_K + R_M. \tag{8.57}$$

In (8.57), the symbol S_l stands for the number of inputs of LUTs. In the discussed case, it is enough $S_l = 5$ for satisfy (8.57). If (8.57) if violated, the states $a_m \in A$ should be encoded in a way minimizing the number of LUTs in the circuit of LUTer1.

The table of EMBer includes the columns $K(a_m)$, $Y(a_m)$, m. In the discussed case, this table has 12 rows (Table 8.13).

Table 8.12 Table of LUTer1 of Moore XFSM $PY_{XC}(\Gamma_{17})$

a_m	$K(a_m)$	$CS(a_m)$	Z_m	m
a_1	000**	0000	–	1
a_2	001**	0001	z_4	2
a_3	010**	0010	z_3	3
a_4	01101	0011	$z_3 z_4$	4
a_5	01110	0100	z_2	5
a_6	01100	0101	$z_2 z_4$	6
a_7	100**	0110	$z_2 z_3$	7
a_8	101**	0111	$z_2 z_3 z_4$	8
a_9	11100	1000	z_1	9
a_{10}	11101	1001	$z_1 z_4$	10
a_{11}	11000	1010	$z_1 z_3$	11
a_{12}	11001	1011	$z_1 z_3 z_4$	12

Table 8.13 Table of EMBer of Moore XFSM $PY_{XC1}(\Gamma_{17})$

$K(EY_q)\, z_1 \dots z_4$	$EY_q\, y_1 \dots y_6$	q
0000	000000	1
0001	110000	2
0010	011000	3
0011	100100	4
0100	110000	5
0101	001010	6
0110	001101	7
0111	100100	8
1000	010010	9
1001	000001	10
1010	100100	11
1011	001000	12

The part of the logic circuit is shown in Fig. 8.27. It represents the block LUTer1 and EMB of Moore XFSM $PY_{XC1}(\Gamma_{16})$.

Let a GSA Γ include T_0 different collections of microoperations (CMO) $Y_t \subset Y$. Let the condition (8.52) be violated. Let us encode the collections $Y_t \subset Y$ by binary codes $K(Y_t)$ having R_Y bits:

$$R_Y = \lceil \log_2 T_0 \rceil. \tag{8.58}$$

Let us use the variables $z_r \in Z$ for the encoding, where $|Z| = R_Y$. Let the following condition take place:

$$2^{R_Y} N \le V_0. \tag{8.59}$$

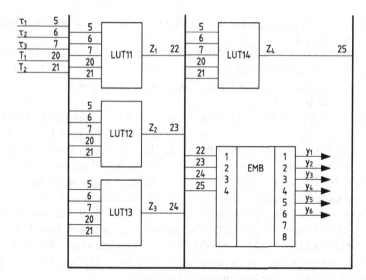

Fig. 8.27 Logic circuit of LUTer1 and EMBer for Moore XFSM $PY_{XC1}(\Gamma_{17})$

Fig. 8.28 Structural diagram of PY_{XC2} Moore FSM

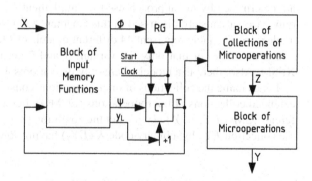

In this case the model of PY_{XC2} Moore XFSM is proposed (Fig. 8.28).

In this model a block of collections of microoperations implements the system (8.54). The difference between XFSMs PY_{XC1} and PY_{XC1} is the following. In XFSM PY_{XC1} the variables $z_1 \in Z$ encode the states of XFSM. In XFSM PY_{XC2} the variables $z_r \in Z$ encode the collections of microoperations. Obviously, the LUTer is used for implementing BIMF, the LUTer1 for implementing BCM, and the EMBer for implementing BMO.

The design method for Moore XFSM PY_{XC2} includes the same steps as the design method for Moor XFSM PY_{XC1}.

Let us discuss an example of design for Moore FSM $PY_{CX2}(\Gamma_{18})$. Let the GSA Γ_{18} have the same structure as the GSA Γ_{15} (Fig. 8.2). But let its operational vertices include the following collections of microoperations: $Y(a_1) = Y_1 = \emptyset$, $Y(a_2) = Y(a_4) = Y(a_{10}) = \{y_1, y_2\} = Y_2$, $Y(a_3) = Y(a_8) = \{y_3, y_4\} = Y_3$, $Y(a_5) = \{y_1, y_4, y_5\} = Y_4$, $Y(a_6) = Y(a_{11}) = \{y_4, y_6\} = Y_5$, $Y(a_9) = \{y_3\} = Y_6$ and

$Y(a_7) = Y(a_{12}) = \{y_3, y_5, y_6\} = Y_7$. So, there is $T_0 = 7$; therefore, there is $R_Y = 3$ and $Z = \{z_1, z_2, z_3\}$.

Let the FPGA in use include EMBs having configurations 16×4 and 8×8 (bits). It means that the condition (8.52) is violated because of $2^R \times B = 16 \times 6 = 96 > 64$. So, both models PY_{XC} and PY_{XC1} cannot be applied. But the, model of XFSM $PY_{XC2}(\Gamma_{18})$ can be used.

In the case of FSM $PY_{XC2}(\Gamma_j)$, the step 6^a is the encoding of CMOs $Y_t \subset Y$. Let us encode the collections of microoperations in the trivial way: $K(Y_1) = 000, \ldots, K(Y_7) = 110$. Now the table of LUTer1 can be constructed. This table includes the columns a_m, $K(a_m)$ $K(Y_t)$, Z_m, m. The only difference between tables of LUTer1 for $PY_{XC1}(\Gamma_j)$ and $PY_{XC2}(\Gamma_j)$ is reduced to the meaning of the third column of tables. In the case of PY_{XC2} this table includes a code of the collection of microoperations $Y(a_m)$ generated in the state $a_m \in A$. In the discussed case, this table has $M = 12$ rows (Table 8.14). The codes $K(a_m)$ are taken from Fig. 8.23.

The table of EMBer for Moore FSM $PY_{XC2}(\Gamma_j)$ includes the columns $K(Y_t)$, Y_t, t. In the discussed case it is represented by Table 8.15. The last row of this table corresponds to "don't care" assignment of variables $z_r \in Z$.

Let us point out that each of discussed models has four variants. The variants are determined by the approach used for implementing the function y_L. As a rule, only base variants are discussed in this Chapter. All possible variants are shown in Table 8.16. The table represent 44 different models of Moore XFSMs.

Let us point out that only the variants 1 and 2 are possible for PY_{XC2} Moore XFSM. Also, there is a model where only the variants 3 and 4 are possible.

Let us name the collections of microoperations constructed for the set $Y \cup y_{L2}$ as extended collections of microoperations (ECMO). Let a GSA Γ include Q_0 of different ECMO $EY_q \subset Y \cup \{y_{L2}\}$. Let the condition (8.52) be violated. Let us encode the collection EY_q by a binary code $K(EY_q)$ having R_{EY} bits:

Table 8.14 Table of LUTer1 of Moore XFSM $PY_{XC2}(\Gamma_{18})$

a_m	$K(a_m)$	$K(Y_t)$	Z_m	m
a_1	000**	000	–	1
a_2	001**	001	z_3	2
a_3	010**	010	z_2	3
a_4	01101	001	z_3	4
a_5	01110	011	$z_2 z_3$	5
a_6	01100	100	z_1	6
a_7	100**	110	$z_1 z_2$	7
a_8	101**	010	z_2	8
a_9	11100	101	$z_1 z_3$	9
a_{10}	11101	001	z_3	10
a_{11}	11000	100	z_1	11
a_{12}	11001	110	$z_1 z_2$	12

Table 8.15 Table of EMBer of Moore XFSM $PY_{XC2}(\Gamma_{18})$

$K(EY_q)\ z_1\ z_2\ z_3$	$EY_q\ y_1\ldots y_6$	q
000	000000	1
001	110000	2
010	001100	3
011	100110	4
100	010101	5
101	001000	6
110	001011	7
111	010000	8

Table 8.16 Models of Moore XFSMs

Type	Variants			
	1	2	3	4
PY_X P_0Y_X	$\Phi = \Phi(T, X)$ $Y = Y(T)$ $y_L = f(T, x)$	$y_L = f(\Phi, X)$	$y_L = y_{L1} \vee y_{L2}$ $y_{L1} = f(T, X)$ $y_{L2} = f(T)$	$y_L = y_{L1} \vee y_{L2}$ $y_{L1} = f(\Phi, X)$ $y_{L2} = f(T);$
$P_{C1}Y_X$ $P_{C2}Y_X$	$\Phi = \Phi(T, X)$ $Y = Y(T)$ $y_L = f(\tau, X)$ $\tau = \tau(T)$	$y_L = f(\Phi, X)$	$y_L = y_{L1} \vee y_{L2}$ $y_{L1} = f(\tau, X)$ $y_{L2} = f(T)$	$y_L = y_{L1} \vee y_{L2}$ $y_{L1} = f(\Phi, X)$ $y_{L2} = f(T)$
$P_{3C1}Y_X$ $P_{3C2}Y_X$	$\Phi = \Phi(Z, X)$ $Y = Y(T)$ $y_L = f(Z, X)$ $Z = Z(T)$	$y_L = f(\Phi, X)$	$y_L = y_{L1} \vee y_{L2}$ $y_{L1} = f(Z, X)$ $y_{L2} = f(T)$	$y_L = y_{L1} \vee y_{L2}$ $y_{L1} = f(\Phi, X)$ $y_{L2} = f(T)$
$P_{1C4}Y_X$ $P_{3C4}Y_X$	$\Phi = \Phi(Z, X)$ $Y = Y(T)$ $y_L = f(T, X)$ $Z = Z(T)$	$y_L = f(\Phi, X)$	$y_L = y_{L1} \vee y_{L2}$ $y_{L1} = f(T, X)$ $y_{L2} = f(T)$	$y_L = y_{L1} \vee y_{L2}$ $y_{L1} = f(\Phi, X)$ $y_{L2} = f(T)$
PY_{XC}	$\Phi = \Phi(\tau, X)$ $\Psi = \Psi(\tau)$ $y_L = f(\tau, X)$ $Y = Y(T, \tau)$	$y_L = f(\Psi, X)$	$y_L = y_{L1} \vee y_{L2}$ $y_{L1} = f(\tau, X)$ $y_{L2} = f(T, \tau)$	$y_L = y_{L1} \vee y_{L2}$ $y_{L1} = f(\Psi, X)$ $y_{L2} = f(T, \tau)$
PY_{XC1} PY_{XC2}	$\Phi = \Phi(\tau, X)$ $\Psi = \Psi(\tau)$ $y_L = f(\tau, X)$ $Z = Z(\tau, T)$ $Y = Y(Z)$	$y_L = f(\Psi, X)$	$y_L = y_{L1} \vee y_{L2}$ $y_{L1} = f(\tau, X)$ $y_{L2} = f(Z)$	$y_L = y_{L1} \vee y_{L2}$ $y_{L1} = f(\Psi, X)$ $y_{L2} = f(Z)$

Fig. 8.29 Structural diagram
of PY_{XC3} Moore FSM

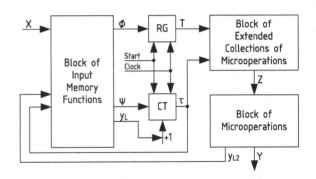

$$R_{EY} = \lceil \log_2 Q_0 \rceil. \tag{8.60}$$

Let us use the variables $z_r \in Z$ for the encoding, where $|Z| = R_{EY}$. Let the following condition take place:

$$2^{R_{EY}}(N + 1) \le V_0. \tag{8.61}$$

In this case the model of PY_{XC3} Moore XFSM can be used (Fig. 8.29).

In this model, the block of extended CMOs implements the system (8.54). Its functions are used for encoding of ECMOs. So, the BMO generates the functions of the system $Y(Z)$ and the function

$$y_{L2} = y_{L2}(Z). \tag{8.62}$$

The variable y_L is generated as a function

$$y_L = y_L(\tau, X, y_{L2}). \tag{8.63}$$

Let us discuss an example of design for Moore FSM $PY_{XC3}(\Gamma_{19})$. Let the GSA Γ_{19} have the same structure as the GSA Γ_{15} (Fig. 8.20). But let the vertices of GSA Γ_{19} include the following collections of microoperations: $Y(a_1) = \emptyset = Y_1$; $Y(a_2) = Y(a_5) = Y(a_8) = \{y_1, y_2\} = Y_2$, $Y(a_3) = Y(a_{10}) = \{y_3, y_5, y_6\} = Y_3$; $Y(a_4) = Y(a_9) = \{y_1, y_5\} = Y_4$; $Y(a_6) = Y(a_{12}) = \{y_2, y_4\} = Y_5$, $Y(a_7) = Y(a_{11}) = \{y_4, y_6\} = Y_6$. There are two unconditional transitions where state codes satisfy to condition (3.26): $\langle a_3, a_6 \rangle$ and $\langle a_6, a_7 \rangle$. It means that the variable y_{L2} should be added into collections $Y(a_3)$ and $Y(a_6)$. It leads to the following set of ECMOs: $EY_q = Y_q (q = 1, \ldots, 6)$, $EY_7 = \{y_{L2}, y_3, y_5, y_6\}$ and $EY_8 = \{y_{L2}, y_2, y_4\}$. Therefore, there are $R_{EY} = 3$, $Z = \{z_1, z_2, z_3\}$.

Let us encode the collections $EY_q (q = 1, \ldots, Q_0)$ by the binary codes: $K(EY_1) = 000, \ldots, K(EY_8) = 111$. Now the tables of LUTer1 (Table 8.17) and EMBer (Table 8.18) can be constructed. Let us pont out that structural diagrams are identical for XFSMs PY_{XC2} and PY_{XC3}. It means that blocks of CMO and ECMO are implemented by LUTer1, whereas the block BMO by EMBer.

Table 8.17 Table of LUTer1 of Moore XFSM $PY_{XC3}(\Gamma_{19})$

a_m	$K(a_m)$	$K(EY_q)$	Z_m	m
a_1	000**	000	–	1
a_2	001**	001	z_3	2
a_3	010**	110	z_1z_2	3
a_4	01101	011	z_2z_3	4
a_5	01110	001	z_3	5
a_6	01100	111	$z_1z_2z_3$	6
a_7	100**	101	z_1z_3	7
a_8	101**	001	z_3	8
a_9	11100	011	z_2z_3	9
a_{10}	11101	010	z_2	10
a_{11}	11000	101	z_1z_3	11
a_{12}	11001	100	z_1	12

Table 8.18 Table of EMBer of Moore XFSM $PY_{XC3}(\Gamma_{19})$

$K(EY_q)\ z_1\ z_2\ z_3$	$EY_q\ y_1 \ldots y_6\ y_{L2}$	q
000	000000 0	1
001	110000 0	2
010	001011 0	3
011	100010 0	4
100	010100 0	5
101	000101 0	6
110	001011 1	7
111	010100 1	8

Table 8.19 Models PY_{XC2} and PY_{XC3}

Type	Variants			
	1	2	3	4
PY_{XC2}	$\Phi = \Phi(\tau, X);$ $\Psi = \Psi(\tau, X);$ $y_L = f(\tau, x);$ $Z = Z(T, \tau);$ $Y = Y(Z)$	$y_L = f(\Phi, \tau)$	–	–
PY_{XC3}	–	–	$y_L = y_{L1} \vee y_{L2};$ $y_{L1} = f(\tau, X);$ $y_{L2} = f(\tau, T);$	$y_L = y_{L1} \vee y_{L2};$ $y_{L1} = f(\Phi, X);$ $y_{L2} = f(\tau, T);$

Now, the table for models PY_{CS2} and PY_{CS3} could be constructed. As follows from Table 8.19, the model PY_{CS2} has only the variants 1 and 2, whereas the model PY_{XC3} the variants 3 and 4. Now, all 48 different models of Moore XFSMs are listed.

References

1. Barkalov, A., Titarenko, L.: Logic Synthesis for Compositional Mocroprogram Control Units. Lecture Notes in Electrical Engineering, vol. 53. Springer, Berlin (2008)
2. Barkalov, A., Titarenko, L.: Logic Synthesis for FSM-Based Control Units. Lecture Notes in Electrical Engineering, vol. 53. Springer, Berlin (2009)
3. McCluskey, J.: Logic Design Principles. Prentice Hall, New York (1986)
4. Yang, S.: Logic synthesis and optimization benchmarks user guide. Technical Report, Micro-electronic Center of North Carolina (1991)

Conclusion

Now we are witnesses of the intensive development of design methods targeting FPGA-based circuits and systems. The complexity of digital systems to de designed increases drastically, as well as the complexity of FPGA chips used for the design. The up-to-day FPGAs include up to seven billions of transistors and it is not a limit. Development of digital systems with such complex logic elements is impossible without application of hardware description languages, computer-aided design tools and design libraries. But even the application of all these tools does not guarantee that some competitive product will be designed for appropriate time-to-market. To solve this problem, a designer should know not only CAD tools, but the design and optimization methods, too. It is especially important in case of such irregular devices as control units. Because of irregularity, their logic circuits are implemented without using of the standard library cells; only LUTs and EMBs (and PLAs) of a particular FPGA chip can be used in FSM logic circuit design. In this case, the knowledge and experience of a designer become a crucial factor of the success. Many experiments conducted with use of standard industrial packages show that outcomes of their operation are, especially in case of complex control units design, too far from optimal. Thus, it is necessary to develop own program tools oriented on FSM optimization and use them together with industrial packages. This problem cannot be solved without fundamental knowledge in the area of logic synthesis. Besides, to be able to develop new design and optimization methods, a designer should know the existed methods. We think that new FSM models and design methods proposed in our book will help in solution of this very important problem. We hope that our book will be useful for the designers of digital systems and scholars developing synthesis and optimization methods targeting implementation FPGA-based logic circuits of finite state machines.

© Springer International Publishing AG 2018
A. Barkalov et al., *Logic Synthesis for Finite State Machines Based on Linear Chains of States*, Studies in Systems, Decision and Control 113,
DOI 10.1007/978-3-319-59837-6

Index

A

Addressing of microinstructions
  combined, 36
  compulsory, 36
  natural, 36
Application-Specific Integrated Circuit (ASIC), 27

B

Block of
  code transformer, 67, 212
  collections of microoperations, 215
  input memory functions, 14, 51, 187
  inputs transformer, 161
  microoperations, 14, 37, 38, 51, 62, 95, 101, 187
  output transformation, 200, 201, 203
  replacement of logical conditions, 88
  transformation of microoperations into steps, 107
  transformation of steps into microoperations, 107
Boolean
  equation, 42, 97, 102, 117, 173
  function, 16, 19
  space, 39, 63, 68
  system, 42, 44, 117
  variable, 14

C

Class of
  compatible microoperations, 107, 109
  pseudoequvalent

ELCS, 48
NLCS, 48
outputs, 199
states, 39
ULCS, 47
XLCS, 49
Code of
  chain, 121, 132, 134, 138, 143, 153, 168, 170
  class, 39, 67, 70, 74, 76, 80, 90, 103, 121, 122, 124, 126, 132, 141, 142, 153, 155, 162, 168, 172, 174, 187, 197, 206
  collection of microoperations, 67, 112, 116, 117
  component, 54, 121, 155
  microoperation, 110
  output, 116
  state, 132
Code sharing, 143
Code transformer, 103
Collection of microoperations, 8
Combinational circuit, 12
Computer aided design, 1, 21
Configurable logic block, 16
Control
  memory, 35
  unit, 1, 2
Counter, 38

D

Data-path, 1
Decomposition
  functional, 26, 27, 145
  structural, 106

© Springer International Publishing AG 2018
A. Barkalov et al., *Logic Synthesis for Finite State Machines Based on Linear Chains of States*, Studies in Systems, Decision and Control 113, DOI 10.1007/978-3-319-59837-6

Printed in the United States
By Bookmasters